Colkirk Tales

Crumps Barn Studio
Crumps Barn, Syde, Cheltenham GL53 9PN
www.crumpsbarnstudio.co.uk

Copyright © the estate of Alfred Absolon 2022

The right of Alfred Absolon to be identified as the author of this work has been asserted by him in accordance with the Copyright, Designs and Patents Act 1988.

All rights reserved. No part of this publication may be reproduced, stored in a retrieval system, or transmitted in any form or by any means, electronic, mechanical, photocopying, recording or otherwise, without the prior permission of the copyright owner.

Cover design and map by Lorna Gray
Illustrations and photographs copyright © the estate of Alfred Absolon
All illustrations by the author, except *The Cottage* by Peter Absolon,
after an original drawing by the author

Inside front cover: *Alfred sketching, 1920s*
Inside back cover: *The cows coming in for milking at the back of the Cottage*,
painted by the author *c.*1910

Printed in Gloucestershire on FSC certified paper by SevernPrint,
a carbon neutral company

ISBN 978-1-915067-15-9

Colkirk Tales

ALFRED ABSOLON

A memoir

Crumps Barn Studio

INTRODUCTION

These are tales of a forgotten life in days gone by. It's also a story of Colkirk, which is a small village three miles South of Fakenham, in the heart of North Norfolk. The author, Alfred George Absolon (1893–1979) was born in London, but spent much of his childhood in Colkirk. His mother was Amy Chambers, whose parents farmed at Great Ryburgh before moving to The Cottage, a small farm in Colkirk around 1878.

The Chambers family also farmed for many years at the Hall, a large farm in Colkirk and had bought The Cottage many years earlier as a dowager property for Anna Chambers, Alfred's great-great grandmother.

In Alfred's time The Cottage was a small farm of 46 acres, managed by his Aunt Kate; and this is where his memoir is set.

Before the arrival of the railways, life outside the confines of the village was barely known. Even in Alfred's time, everyone in the village was dependent on each other; and on the success of the village's farms to provide their livelihoods and their sustenance. Self-sufficiency was rooted deeply in the collective consciousness. In this, folklore was every bit as important as science.

The experiences Alfred writes about here had a huge impact on him, and it is clear that that he came to see Colkirk as his place of origin. He felt more connected to the Chambers in Norfolk, as farmers, than anywhere else.

When the Great War came, he volunteered with his brother Leonard, and fought at Gallipoli and in Mesopotamia, where he was awarded the Military Cross. When he returned to Colkirk in 1919, it was natural for him to start farming for himself. He and his parents moved to The Grange in School Road, Colkirk, (now called Manor Farm) and until 1927 he farmed 47 acres made up of pightles (small enclosures) and strips of land around Colkirk.

A wood carving by Alfred can be found near the pulpit in St Mary's Church, Colkirk. Also in the Church is a roll call of those from Colkirk who enlisted in the First World War where the names of Alfred and his brother are recorded. In the churchyard are the graves of some of the people mentioned in this book, including Aunt

Kate and others of Alfred's ancestors from the Chambers family.

In later life he felt it important to make a record of a way of life that has now disappeared from the Norfolk countryside forever. This book is his description of that village, its inhabitants and their daily lives at the beginning of the 20th century – an impressive and invaluable archive.

In his words:

> *By writing down my memories of the very last period in the history of rural England, I am trying to preserve at least a little of those times for my sons, for their children, and anyone else who may want to know something of the soil whence they sprung.*

Between 1968 and 1972 he wrote these Tales.

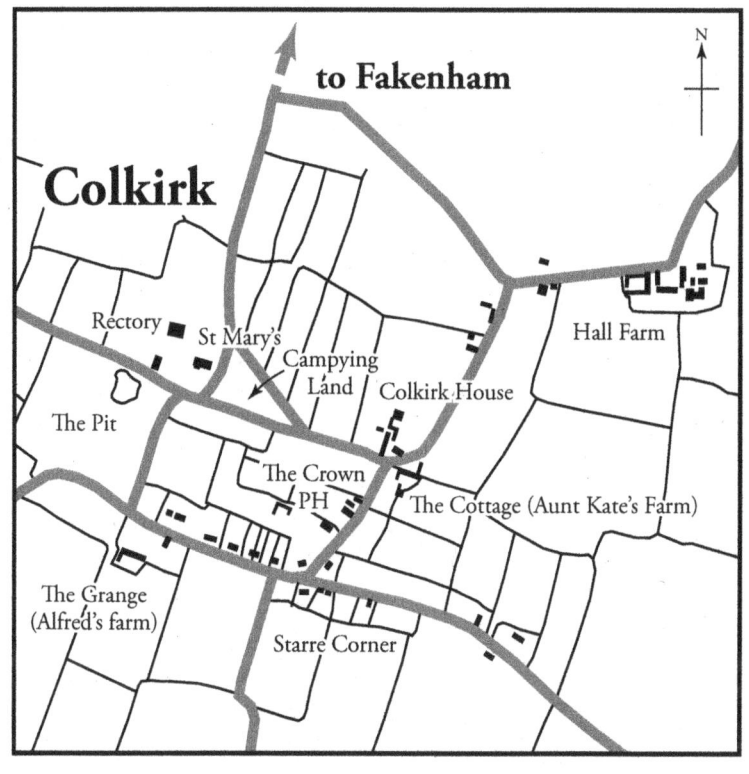

Map of Colkirk village c.1897 (not to scale)

CHAPTER ONE

A NORFOLK VILLAGE

The Cottage

My earliest recollection of Colkirk was, I think, the Diamond Jubilee of good Queen Victoria in 1897. It was an occasion for public rejoicing and a good excuse for a real village do.

The old rector and his energetic daughters were probably at the bottom of it. They were always the prime movers in any village happening, whether it be to celebrate a birth, a death, a marriage, or merely the end of a distant war. Those were the days of real isolation. Any event that did not take place within a few miles hardly disturbed the placidity of life.

I was aged about four, and that year there had been a good and early harvest, the weather was still fine, and the time was ripe for rejoicing. Races and athletic events, both for adults and for children, were held in a big meadow known as Watering Close. This was also, when occasion demanded, the Cricket Meadow. It was owned by my grandfather and used for grazing the cows, who were not disturbed by a mere cricket match. Fielders were expected to shoo them off if they wandered onto the pitch. It seemed most remarkable to me in later years when I visited the Oval that one spectator moving in front of a sight screen was enough to stop a game. We did not mind.

A great pit took a bite out of the meadow. It was full of water and fairly deep on the meadow side. Here, where the bank was conveniently steep, a pole was fixed horizontally over the water with a box at the end of it. The pole was well greased and the box contained, instead of the traditional live pig, a leg of pork. The lads of the village were urged by excited spectators to walk the greasy pole and claim the prize.

One by one they floundered and fell off into the water with a mighty splash. The winner reached the box, made a grab and overbalanced at the last moment. But he did touch it, so he was awarded the prize.

There were dark hints by the losers that he had sanded the soles of his feet.

Church Pit

As the day was ending all repaired to the neighbouring barn, which had been swept and garnished for a feast. Cold roast beef and pickles, beer, plum puddings, and more beer. The plum puddings were boiled in the coppers of the farmhouse nearby – The Grange, now called Manor Farm, in School Road, Colkirk – the house which years afterwards became my parents' home; the house I came home to from the First World War; and the house to which I eventually brought my bride.

Colkirk did not change much during those early years, though the toll of young men who did not come back was heavy. Many of the boys with whom I played cricket and football 'went west', and those who returned were not the same. Jack Ram lost an eye and an arm. He became the village postman and rode into the nearest town on a pony for the letters – a great advance, for before the war the post did not arrive until about six o'clock in the evening.

On a wet cold night, the postman was sure of a glass of whisky when he called at my grandmother's house. So it was that I heard an eye-witness account of the local ghost. Jack Ram told me how he had been riding down Market Lane early one morning for the post, while it was still dark, and his pony had shied at a misty figure in the road. Both Jack and the pony were terrified. It was gone when the pair of them, pony and rider, at last nerved themselves to pass the spot.

One of my aunts who lived at the bottom of the lane had a similar experience. She had visited the Rectory for a game of cards and on the way back caught a glimpse of a misty and, she said, terrifying figure. She fled back to the Rectory in horror and had to be revived and escorted home by the Rector. That was long before Jack's adventure, and Aunt Susan's story as far as we knew was known only to the family. A mystery – and so it remained.

Another strange phenomenon was observed more publicly during a service at the little church of Gateley, near Colkirk. A ball of fire was seen to enter the church door, trundle down the aisle and explode against the altar. The Devil, of course. But perhaps there is a more ordinary explanation for this one – we boys had also been scared by a ball of fire which came through the open door of the barn when we were playing inside on a rainy day. It passed quite slowly over our heads and hit a scythe that was hanging on the wall with a great clang.

Vagrant lightning?

But then Gateley Church rather went in for devils. One of the old pictures painted on the Chancel screen showed a fine horned devil being conjured out of a jackboot: the origin of the painting was probably a squire with a bunion and a clever cobbler. But even in our young days the Devil was a very real person and well respected. I remember while being prepared for confirmation the old Rector asking me 'Is the Devil omnipotent?' Speaking from experience I replied 'Yes!' But that was the wrong answer.

I was sent to boarding school at the nearby market town of Fakenham when I was seven years old, which was about 1900. It was a girls' school really, but a number of small boys were taken as well – only one other boarded besides me, until my younger brother joined me later, and a few day boys.

We were weekly boarders, and how we longed for Saturday when my Aunt would arrive in the pony and trap to take us home for the weekend. We spent all our weekends and short holidays at our grandmother's farm, and often the long summer holidays too.

At The Cottage, we learnt to work hard and play hard and acquired the lore and customs of the country. We learnt to handle and tend animals, collect eggs, and help with all the little jobs that need doing on a farm. In due course I was taught to load a wagon, starting with hedgerow cuttings – awkward stuff to load, slippery and springy.

We were carting them off to the stack yard one day to make stack bottoms for haysel or harvest, when I fell off, head first making a nice little round hole with my head in the soft earth. I wasn't hurt much. Nobody bothered. That was the way in the country. Somebody felt you to see if any bones were broken and if not, all right, get on with the work – much more important than a small boy's aches and pains.

The next event was rather more serious. It happened at harvest time and I was helping on the stack. Missing a sheave (a 'shove' in Norfolk) I stuck the fork into my foot. Luckily I had good thick boots on, but when the fork was pulled out the blood spurted up and I was carried down the ladder and into the house, not feeling too good. It was good being carried around in a bath chair, but I was back at work in the fields by the end of harvest.

Old Bartaby, one of the farm men, carried me down the ladder – in his younger days he had worked on the Hall Farm in Colkirk, which belonged to my great Uncle. But by the time I knew him he had 'took religious'. He would suddenly turn to you and say, 'Are you saved?' Most disconcerting.

My great Uncle, Thomas Chambers, had a large farm of 700 acres at The Hall. He was a man before his time and had invented all sorts of farm machinery. Bartaby used to help him, and sometimes he worked all night making parts for my Uncle's inventions.

Up to my boyhood, every large farm had its own

forge and carpenter's shop. My Uncle had little railway lines laid down from the 'turnip house' to the fat bullock boxes and yards, so that one man could push a whole truckload of feed to the stock in one go, instead of carrying it a skep at a time. So automation is not so new.

He invented a method of drying out a hay or corn stack by blowing hot air into it, a practice that was hailed as a new idea in the Second World War. He also devised a means of ploughing a field by means of cables worked by a portable engine that was set in the corner of a field. The 'portable' was a single piston engine, mounted on ordinary wooden wagon wheels and drawn by a couple of farm horses. It was still working within my time.

My grandmother's farm was attached to The Cottage in Colkirk. It was not a large farm and we usually got to know the men who worked on it quite well. The first I remember was Wake, who grew a large black beard and a numerous brood of children – how they all packed into their tiny cottage is a mystery. In fact I think that was why Wake left.

By modern standards those cottages were a disgrace, tacked onto the stables and two of them having no back door. Each cottage had one fairly large room downstairs and two small rooms above, one leading into the other – no scullery, no pantry, no drains, but a good large shed attached. The shed was a must in the country.

Burton came next – a great character who had migrated to London in his youth, to work in a dairy. Cows were kept in the middle of the city in those days;

hand fed all the time, never seeing a meadow or a stream. They were mere machines for producing milk, and not very good milk at that. It was distributed in open cans and bailed out with a measure on dusty doorsteps. No wonder the infant mortality rate was high. And I'm afraid our milk in Colkirk was little better, carried in open wooden buckets from the cowshed and poured into open pans.

The milk that was unsold remained in the pans for the cream to rise – next day it would be skimmed off and poured into large crocks to ripen for butter. People came to buy their milk at the door, bringing their own jugs and cans, and it was bailed out to them with pint or half pint measures. No one bothered much about washing hands, which were constantly in contact with the milk in bailing it out. I have seen things in cowshed and dairy which would make a modern sanitary inspector's hair curl. Nobody thought a thing of it then.

I remember one day an old lady from the village came to the door for milk and she produced from her jug a gorgeous apple, which she gave me. We were free of all the apples of the garden, but that apple tasted better than any of them, or indeed than any apple I have tasted since. Perhaps it was the kindness and love with which it was given.

The old lady was the wife of the village carrier, a Crimea veteran with a peg leg, which was due not to the war but to a wagon accident. He went into the market town every day shopping for people, and on his pony cart he had room for one passenger.

He must have had a small army pension, for the amount he earned this way could not have brought him a living. Later I learnt it was the Rector who set him up in his little business. That was how things were done then. There was no National Insurance, no National Assistance, and no old age pension – only the dreaded workhouse.

So people helped each other. The better off felt it their duty to help those in distress. And I don't think the recipients felt that they were objects of charity – it was simply the expression of a relationship between the people of a remote parish – a 'backward' one, even sixty-five years ago.

I have read that the same sort of feeling exists even today in certain slum areas of the large cities. But I cannot help feeling that we have lost something in these days of the welfare state. I remember another old man, long past work, who came to the door regularly for his pay. He used to be a mole catcher and although everybody knew he hadn't caught a mole for years, the polite fiction was kept up until the day of his death. Today he would be a Pest Officer, a Government Official, and would be marched off his fields on the appointed day to lose his identity in the ranks of the old age pensioners.

Old Burton was a true countryman, in spite of the London experience. He always wore a faded plaid scarf, a woollen one, knotted about his throat. I never saw him without it. Every autumn he killed a pig and salted it down in a barrel which stood at the corner of the kitchen. He lived on that all through the winter months and towards spring the smell in that kitchen was 'enough to knock yer backards'. His poor old wife could not eat it and would have fared badly but for the ministrations of my Aunt.

Mrs Burton claimed to have second sight. She said she often saw people walking about with strange clothes on. And she prophesied that my brother also would see 'things' because he was born in 'chime hours'. Certainly one night he claimed to have seen a man in our bedroom, dressed in strange clothes which from his description I should now recognize to have belonged to the eighteenth century.

Mrs Burton gave out that she was a witch, but a white one. This meant not only that she could see things, but she could cure warts and she understood the use of wild herbs. We were never afraid of her; in fact we were very fond of the old dear. But there was a black witch in the village – of her I shall tell later.

Before the Burtons came to us at The Cottage, they acted as caretakers at the Hall, which was empty at the time, the land being farmed by an absent tenant. Burton kept the house aired and Mr Burton acted as yardsman. My great uncle Thomas Chambers had departed this life, but he had left part of it behind him, as it turned out.

My aunt used to visit Mrs Burton at the Hall with a basket of comforts, and occasionally she would take me with her. One entered the kitchen, I remember, through a large, stone-floored 'back kitchen'. There would be a large wood fire burning in the open hearth and a worn old rug on the stone floor in front of it.

Mrs Burton once told us 'While I'm asettin hare a great ol dorg come in and set hisself down by the fire. T doan't take no notice o him an he doen't take no notice o me. Presently he get up and go out again.'

My Aunt said she could remember a dog there in her childhood very like the ghostly one Mrs Burton described.

Old Burton said another time that he often saw as he went out in the grey dawn the figure of a man dressed in a tailcoat with brass buttons at the back, standing in the corner rubbing his hands. He could never understand what he was doing. Then my Aunt remembered that when my great uncle Thomas was alive a basin stood there and he would pull off his jackboots and wash his hands before going into the house. Neither of the Burtons was at all frightened. They seemed to look on these things as quite normal and they came from a village on the other side of the county, so they knew nothing of our family history.

When a man left a farmer's employ in those days, it was the understood thing that he should be allowed to borrow horses and wagon to cart his family and furniture to his next employer. This is what in due course the Burtons did. Old Burton always said that he would come into some money. Nobody believed him, but we heard after he left that he actually had. But poor old Mrs Burton had died before then.

About this time, when we were still at school in the

local town, the whole school was taken on an outing to see the men come home from the Boer war. They disentrained at the railway station and marched through the town, a brave sight in their khaki and slouch hats.

Little did we think then that we should ourselves ride out of the town only a few years later as members of a troop of Norfolk Yeomanry. At the end of two World Wars, we were demobilised singly and sneaked home, almost as if it were a shameful thing to have served.

After the last war in fact I was told, or it was inferred, more than once, that I was lucky to have been in the army away from the bombing.

Having served at home on the east and south coasts, taken part in a landing, and survived a bombing and torpedoed ship, that did seem a bit thick. But these remarks were not made in our beloved Colkirk, which I left long before the last war, and what a change! I went into the Crown, Colkirk's only pub, and I found the old chaps I knew huddled in one corner. It used to be a quiet pub where a man could get a drink and a game of dominoes away from his family in the evening. No woman was ever seen there then.

The land which I had known and cultivated was sold and largely covered by a rash of ugly little red brick bungalows, inhabited by incomers who had been brought there by wartime industry and airfields. The old cricket meadow was desecrated by chicken runs. But I have jumped a few decades. This story was meant to describe an old-time village, inhabited by real live people, the salt of the earth. I have given you a glimpse of the corpse.

CHAPTER TWO

IONA

The snare

Leglin is the next man I remember on the farm. He was clever with tools and showed a lot of initiative about the yards and buildings but was bad with stock. He used to have sudden and inexplicable fits of cruelty, and it was because of this cruelty that he had to go. He went away from Colkirk for a time, and then came back to a job on another farm.

On his return we learnt that he had been 'converted' by the Salvation Army, those good and earnest people, but unfortunately it did not last. Once back, Leglin started a religion all his own, which degenerated into wild orgies and dances, creating such a scandal in the village that the police had to be called in to deal with the situation.

The Law in Colkirk was represented by a policeman who lived in another village some three miles away. He was rarely seen, yet seemed to know everything that was going on. In the few cases of chicken stealing – about the most serious crime committed in his district – he would go straight to the shed of the culprit and there find the corpse of his victim, or at least sufficient feathers and remains to convict him of the theft.

The resulting fine was probably paid by the Rector, after a private interview in the study. Juvenile crime was dealt with summarily by the policeman himself or by the school managers. He turned a blind eye on the poaching of a few rabbits and hares, although undoubtedly he would know about it, but he would go for any organised gang from the town. That was the way of country policemen – they saw beyond the letter of the law and applied it with a broad countryman's instinct for what was appropriate.

As we grew older, we learnt how to set snares for rabbits and hares, but that was on our own land. Callous though I was in those days, after I had heard a hare caught in one of my snares on a still night I set no more. On the whole, though, I went along with the toughness those countrymen had to have in their dealings with animals.

They were not deliberately cruel (Leglin was the exception, for he was what I suppose would now be called a psychopath). People lived very close to nature, which is itself cruel. Today, if we were asked to kill our own meat, possibly many of us would turn vegetarian. Countrymen I knew often had to kill their own meat, animals they had reared and cared for.

Most cottages had a pigsty in the garden and in it a pig – probably the most prized possession of the family. When the time came for it to be killed, the family turned to and dealt with every possible part that could be eaten. We are so refined now that we pay people to do these jobs. Then it was the normal thing.

When I was a small child I was taken by a nursemaid, as a great treat, to see a pig killed. It scared me so much that ever afterwards I tried to avoid the ceremony. Later on I would take my own fat pigs down to the butcher or put them in the market for sale. But the result was the same. One became callous – animals were there to be killed for food, and sometimes for sport. I well remember my first effort in *that* direction. I was given some salt in a screw of paper and told to go round the stack yard and catch the sparrows by putting salt on their tails!

Later we were taught how to make brick traps and sieve traps for small birds, and how to go round the stacks at night with a sieve stuck on a two-tined fork, clapping it on to the side of the stack where the birds were roosting.

The man who took Leglin's place was Knights. I remember driving over to a remote farm on the Wells road to engage him – a tall, handsome man whom we all took to straight away. Knights was better educated than any of the others, the son of a small farmer from the other side of the county. He had at one time been manservant to a gentleman who travelled in Europe, so he had a lot of interesting stories to tell us youngsters.

He also taught me the little I know about farming and the care of stock, which stood me in good stead later. One never finishes learning about the land, but a good practical grounding is essential. We were old enough by the time Knights came to do really useful jobs and he would set us to work and see that we did them properly.

About this time we were taken away from the little school in Fakenham and sent to a boys' boarding school, dignified with the high-sounding title of Malden College. However, we still spent the long holidays in Colkirk and then we would quickly slip back into the old ways of country life, forgetting the trim playing fields, the swimming bath, the drill sergeant who taught us to box, and the rudimentary army parade and field exercises. Anyway, it didn't last very long. My father got into financial difficulties and back we went to the farm, for he could no longer afford our school fees. That suited us very well. We did our best to earn our keep, and I never went back to school.

Knights was a grand chap and worthy of a better job except for one failing – every now and again he would go off and get blind drunk. Whisky was cheap in those glorious days and once Knights got hold of a bottle he would go on drinking until it was finished. But he never allowed it to interfere with his work and he never neglected the animals or was cruel to them.

I came across him several times, lying helplessly sozzled in the barn. His lapses were only occasional and I think they were brought on by fits of desperate depression. Modern psychologists would have a name for it. To us it was just a weakness to be forgiven.

I said that he was educated, but not at all in the sense of a modern state school. He combined the native shrewdness of his ancestors with the ability to apply and adapt what he had read. He was good at doctoring animals and supplemented the animal lore he had learnt from his father with a modern veterinary book. He always kept a dried puffball in the barn to dress a wounded animal – and the cuts he himself received in the course of his work. I myself have used it on a cut or a scratch – shaking the powder from the puffball on to the wound immediately stopped the flow of blood and formed a scab, effectively keeping out infection.

Some of his remedies were not quite so simple, as when he crossed swords with the 'Black Witch' of the village.

It was before the days of motor cars, at least in our part of the country. At milking time, to get the cows home, one simply wandered out to the meadow where they happened to be that day and opened the gate, making appropriate noises. Then one stood back to admire the scenery or have a chat with a passer-by while the cows, knowing their way home, wandered through the village in their own time, and each to her own stall. Cows, incidentally, have a very strong sense of individuality, and it is part of successful management to recognise this.

A young cow is taught her manners by her elder sisters in an unmistakeable way. On one occasion, however, one of our young cows forgot herself and wandered through the open gate of a cottage garden, enticed by some very juicy looking cabbages. No good countryman ever leaves his gate open. Unfortunately this garden belonged to the 'Black Witch', with dire results to the cow. When she got back to her stall she went completely wild and unmanageable and generally behaved as no self-respecting cow should.

Knights said 'That old Warmint has bewitched her. I know what to do.' So he pulled a hair from the cow's tail and nicked the cow's ear, got a drop of blood from the ear and soaked the hair in the blood. Then he most solemnly carried it into the kitchen and burnt it on the kitchen fire.

'Ah,' he said, 'now the old so and so is going through it. She'll have to take off the spell.'

Sure enough the cow immediately became quiet again, letting her milk as a good cow should. The underlying idea, as Knights explained to me, is that as the blood-soaked hair burns, the witch undergoes the pains of burning and has to take the spell off to obtain relief. A simple explanation, allowing for coincidence, might be that the cow was suffering from a surfeit of cabbage. Nevertheless, a few generations back the witch herself might have been burnt.

I asked Knights how he knew the way to break the spell. He told me:

> When I was a boy my father had trouble with his pigs. They were dying one by one. He had unfortunately fallen out with a neighbour, known to be a male witch, who had cursed him and all that was his.
>
> My father said one day 'I know how to break the spell!' He killed one of the ailing pigs – this was a very strong spell – and, catching some of its blood in a saucepan, he mixed it with some hair from the pig's back. The he carried this brew into the house and put it on the kitchen fire to boil.
>
> 'Now boy,' he said, 'shut and bolt all the doors and windows. Don't you let him in on any account. He'll have to come here to take the curse off. He'll be in torment.' Presently there was a wild shouting and banging on the door.
>
> 'Let me in! Let me in!' the witch cried.
>
> But my father firmly told me 'Don't you let him in. He's going through it now. He'll have ter take the curse off.' After a bit the shouts and screams died down into low murmurings and at last there was peace. The rest of the pigs got better.

Mind you, I could produce a like condition in pigs by putting just a few crystals of soda in their feed. Soda is death to pigs. Our maids, who used soda for washing up, had strict orders never to put washing-up water in the swill pail. A careless maid did it once and we lost a good sow. We did a post-mortem – at least Knights did, while I watched – and the stomach was all inflamed.

Pigs are delightful animals and one can learn a lot by watching them. They are naturally clean creatures, only humans make them dirty. True, they love to wallow in mud, but this is the way they cleanse themselves of vermin. No doubt they feel just as good after a good wallow as we do after a good bath. They like to root in the open after grubs and fungi. A sow rooting in her sty probably needs minerals – a shovelful of small coal remedies that.

Many people take away the straw from a sow at farrowing time but this is a mistake. Once, shortly after I took charge of the farm myself, I watched a sow loose in a meadow spend all day collecting wet grass and weeds to make a bed for herself to farrow in. After that I always gave my sows plenty of short straw at farrowing time. They would spend hours happily making a bed, exactly to fit their bodies, about two feet high and lower at the head than at the rear.

When the time comes, the sow will lie down with contented grunts and deliver her pigs without any trouble. Even before they are properly clear they will scramble to the milk bar and start sucking away.

Another thing I learnt was not to let them see me watching. No animals like to be watched at such a time. I once had a mare in foal. She was called Iona and I prized her very much, for she was a pedigree Suffolk Punch.

The vet told me she was 'not right' and I should be lucky to get the foal alive, so I sat up with her night after night. At last a day dawned when it was obvious that this was going to be the day. So I stayed with her all day long, equipped with scissors, twine, disinfectant and everything else I could think of. Towards evening I slipped into the house for a cup of tea. When I got back to the stable, prepared for the worst, there was the foal, as lively as could be, and Iona, the mare, on her feet feeding him. She had waited until I was out of the way. Birth is a private affair.

I loved my horses. When the place was eventually sold over my head and I could not muster enough money to start elsewhere, the hardest thing of all was having to give them up and sell them. Every evening, no matter what I was doing or what the weather, I would go myself to the stables to rack them up – that is, to fill their hay racks and shake down their bedding. They would whinny as they heard my footsteps approaching. It was the same in the dark mornings of winter, for they had to be fed in good time before starting the day's work. In the summer they were turned out to grass, but they had to be brought in and given a feed just the same. Their hard work would be over by the summer, for the ploughing and sowing were all done and there was only light horse-hoeing, haysel and harvest.

Ploughing was the hardest work, for it meant a steady pull all the time. A plough horse is always in the collar and when they are new to it horses take some time to get used to the strain. A cart needs a good pull to start it rolling, but after that it is easier. A horse that is used to a cart cannot understand why the plough does not do the same thing. Once trained, however, a good plough horse will keep up a steady pace for hours on end.

There is no need to use the reins – the plough horse learns to obey the voice. 'Cupya!' means veer to the left; 'Wheesh!' keep on that course; and 'Wheesh! Wheesh!' move over to the right. These commands are right at any rate for Norfolk. I rather think other counties have different words of command.

Every ploughman takes a pride in driving a straight furrow across a field. A peeled stick is set up as a mark, the plough is specially set and the horses coupled wider than usual. Then you get squared up, get a good sight on your mark stick, perhaps a couple of hundred yards away, and off you go – resisting the impulse to look back and see if there is a wiggle in the intended straight line. The horses know it is a special job, for normally one horse walks in the furrow.

When the horse knows his man and the man his horse, there is a marvellous sympathy between them. Anyone who has jumped a horse knows that. If you are scared, the chances are that your horse knows it and he muffs the jump. As the old saying goes, 'You have to throw your heart over the jump first. Your horse will follow.'

It is much the same when it comes to ploughing with a straight furrow.

Alfred harrowing in the 1920s

The horse is one of the noblest of God's creatures, which makes it particularly painful to see one ill treated. From the house where I lived when I started writing these stories I would sometimes see some fat, heavy-handed rider rolling about in the saddle and pounding along the hard road which leads up to the open Cotswolds. It is a favourite route for hired hacks. I had to restrain myself from rushing out to tell these shabby riders a few things. In fact, one can tell at a glance if a person can ride or not. One or two passing by were a joy to see, but too many seemed to think they were driving a motor car with four legs.

Best of them all, I remember a mare called Meg. Things were looking up for me when I bought her. It was before the bad times. I had a young wife and a son, and I really thought they deserved something better to drive than the old pony that was always lame. So I bought a beautiful little mare and named her Meg, after a mare I had ridden in the army. She was gentle and strong and could do light work on the land. It was between haysel and harvest, so we decided to take a summer holiday – a weekend by the sea at Wells. Meg had only been with us a fortnight and it was a good excuse to try her out.

We drove over to Wells and put up at a little hotel, turning Meg loose in the adjacent paddock, which belonged to the hotel. Next morning my first thought was for the little mare, so off I went down the road to see how she was. She heard my footsteps, cocked her head and came trotting over with a friendly whinny. And I had only had her a fortnight!

Many years later when I went back to Norfolk and looked around the old places in Colkirk I saw an old horse, a Suffolk Punch, grazing in a familiar meadow. Surely I knew that mare?

I nipped over the gate and, corning nearer, recognized Iona, whose foal had been born that teatime when my back was turned. She must have been incredibly old – her poor bones stuck out all over the place. I made much of her, but she did not know me. There were no signs of any ill treatment, just weary old age. I thought of the hours she had worked for me, always willingly, and the hours I had worked for her, dosed her with physic, and slept in the stable; and the joy she had obviously felt over her foal. I thought of the sad day I had parted with her. Just as well she did not remember. When I walked away from that meadow it meant nothing at all – to her.

Pony (Meg) and trap outside The Grange in 1926, with the author's eldest son Eric aged 1 year and 9 months

CHAPTER THREE

TALES OF LONG AGO

The Village Cobbler

Certain smells are powerfully linked with memories. The smell of well-tanned leather never fails to remind me of old Billy Dunn, the village cobbler. We enjoyed taking our boots to be repaired. We liked the place with its comfortable smell and we liked old Billy's salty talk.

He was something of a socialist, as I believe many cobblers are. Is it something to do with the way boots are worn indicating the character of the wearer? Or is it because the cobbler makes a living from the needs of a wage-earning world? Either way, Billy had a sour outlook on established things.

He had a tiny cottage, and a small room for his workshop, with a lean-to kitchen and a bedroom above. Not that we ever penetrated that far. Billy was a bachelor; at least, I had never heard of a Mrs Dunn. Perhaps there had been one, which might have explained something of his crabbed attitude to life.

Billy used to sit on the floor with his back to the wall. In front of him was a low bench divided into small compartments which held his nails, tacks, knives, awl and wax. These benches are now museum pieces. Between his knees was his last, and all around him on the floor boots and shoes were heaped in confusion. How he sorted them out for their rightful owners I don't know. From old Billy's talk, you could certainly glean evidence about personality showing in the shoes.

'Them's Mrs B's. She do wear 'em on the outside. Waddles like an ol' duck, she do.'

Around the wall hung bit s of leather, straps, bridles and all sorts of junk. A tiny cast iron fireplace contained the ashes of the last fire in the grate or, if it was a very cold day, there might be a smouldering fire. On a shelf above were his shaving things, for on Saturday nights Billy was also the village barber. His equipment consisted of a piece of mirror, a shaving brush, a fragment of yellow soap and a smoke-blackened tin. A chair with a broken back completed the outfit. Perhaps he also cut hair but I have no evidence of that. Hair-cutting was usually the mother or wives' job.

Billy Dunn's was a great centre for gossip and tales. He would make his comments between hammer blows. 'Them furriners – *bang* – ought to be hung – *bang, bang.*'

People rarely read newspapers, if they could read at all. News filtered from mouth the mouth, and there were always tales of bygone days to while away the time. To village folk, fifty years ago was like yesterday, and just as real. There was continuity to life in the country which took no count of time.

Another Billy Dunn was son of the Rector's coachman and nephew of Billy Dunn the cobbler. Young Billy was a nice lad and very fond of cricket, but not very bright. He left school at fourteen and after a spell as odd job boy at the Rectory, he was sent as boot boy to a big house in a neighbouring village. It was, I suppose, the first step towards footman and butler.

Billy was always high spirited and ready for a bit of fun. One night he played the old country game of frightening a girl by jumping out of a hedge in a dark lane. Instead of catching him a good 'un on the side of the head as most country girls at that time would have done – it was often, in fact, a sort of preliminary to courtship – this girl ran crying to her mother. When

Mum learnt that the miscreant was that new boy at the Hall she duly went there and complained.

The worst that could have happened to Billy was a wigging from his master, or a note to the Rector. But next evening, when he went to play cricket with the other boys on the green, they teased him unmercifully about his escapade, hinting dark things that could happen to him. Suddenly one of the lads called out 'Look out Billy! Hare come the p'liceman!'

There was no policeman, but Billy took fright, snatched up his coat and ran. Everyone thought he had run home. But next morning there was no Billy at the coachman's cottage and people began to get worried. Some said, 'He'll come home when he's tired of hiding!'

But he didn't.

On Sunday in church the Rector announced: 'There will be no sermon today. All men and boys will meet at the Starre Corner in half an hour's time to search for Billy Dunn.'

Notice the 'will' – what the Rector said in those far off days was an order. Besides, it was a fine Sunday and there wasn't much else to do. We boys were naturally delighted.

The expedition was put in the charge of Buckingham, the gamekeeper, as he was used to handling beaters at a shoot. The theory was that Billy had run for home across the fields and had come to some harm on the way. We imagined him lying cold and pallid in some ditch. But we searched all the woods and the fields for miles in that direction and found no trace of him.

Nothing was heard of him for more than a year. Then one day he walked into his home. He had run in the opposite direction, to a little port on the coast, probably the only sea that Billy knew – people at that time seldom had summer holidays, usually they only went for one day. Parties got together and went twelve miles to the beach in farm wagons decorated with green branches, or in wagonettes hired from the town.

Billy was in luck. A little Dutch coaster was lying in the quay, waiting to go out on that day's tide, and he went on board asking for a job. Again he was in luck – they happened to be short handed. He was signed on as deck boy.

What an experience for a country boy, knocking around the ports of northern Europe in a little coaster! Billy could stand the hardships all right. He was used to hardships. He left Norfolk a frightened boy and he came back a man, to be lost in the First World War. So often it seemed that the men we could least afford to lose 'went west'. I could name so many from that one small village.

We used to visit a remarkable old lady in the village by the name of Pamela Howard. We usually brought with us some little gift from our aunt, and later on as we got older we occasionally brought a brace of rabbits we had shot or ferreted. Pamela had been a schoolmistress in her younger days and we were a bit afraid of her.

She was very eccentric and lived in a long, low house with a paddock attached. In this paddock she kept a donkey, surely a pensioner, for we never saw it driven on the roads. Then one day the donkey died. Poor old Pamela was very upset and would not let anyone bury it for her. She stoutly maintained that the donkey was not dead – it was just not feeling very well.

Some wit said to me, 'You'd better go and look. It will be a long time before you see another dead donkey.' Well, we went and had a look at Pamela's donkey, lying in the corner of the paddock near the road. It was already making its presence known and something really had to be done about it. One night some kindly young man went and heaped earth over the corpse where it lay in the paddock. This satisfied Pamela for, as she said, when he felt better and wanted to get up he could push the earth out of the way.

To make up for the lost companionship of the donkey Pamela bought a bicycle, which she never rode. One day when we went to see her she showed us that she could nearly ride round her sitting room table, with one hand on the table to preserve her balance.

The old lady had a wonderful store of clothes, all of them 'period', and she never mixed her outfits. One week it would be early Victorian; the next it would be the bustle period. Her store of clothes would have been a rich harvest for a wardrobe mistress of a repertory company.

Pamela regularly wheeled her bicycle to market. She said it was company and handy to hang her bags on. The tyres were quite flat. Pamela and her bike were really inseparable. Somebody told me years afterwards that she once tried to take it up a narrow stairway to the sanctum of a solicitor she had occasion to visit. In spite of Pamela's Victorian dignity the solicitor's clerk was even more frightened of the explosion that would have followed if he had ushered her into the presence complete with her dilapidated bicycle. He refused to let it pass. But others in the downstairs office rushed forward with promises that they would look after it very carefully, so all was well.

Townsmen often get the idea that country people are stupid. Of course this is a fallacy. They may sometimes be backward in things like reading but their intelligence is applied in ways that a townsman may not grasp at all. I am reminded of this fact when I think of old Lake, whom I employed in my later days at Colkirk. He really was an intelligent old fellow. He taught himself to read when he was grown up, most likely with the help of the Rector's daughters.

Lake had started work as a small boy, crow-scaring. All day he would walk round a newly sown field, swinging a wooden rattle and letting out wild yells to keep the birds away. Even I, in my comparatively short recollection, remember the peculiar 'Whoop!' of the bird scarer.

The inn keeper used to grow barley every year in a piece of land attached to his pub and, being plagued with sparrows because of the nearby houses, he kept guard over his precious corn with an old muzzle loader. Every now and then there would be wild cry, followed by the 'pip-bang' of his old gun – the 'pip' was the percussion cap, which went off an appreciable time before the charge exploded,

Old Lake told me that people were more sociable when he was a boy. They would gather in one another's houses in the evening, often with no light other than the fire. Even a rush light was more than some could afford. If the boys kept quiet and stayed on the outside of the circle they were invariably forgotten, so they would stay and listen to their grandfathers telling tales which their own grandfathers had told.

To countrymen, times and dates were not of much importance beyond the last harvest, and this race memory stretched back hundreds of years. Lake often spoke of events long past as if he knew of them personally. One could identify some of them, such as the Peninsular Wars, by their place in history.

Norfolk was a priest-ridden county before the Reformation. We lived not so far from Walsingham and the religious houses owned much of the land round about. I think it is safe to infer that much of the anti-Church and anti-Popery feeling in the county stemmed from those days.

Lake told me about the steep-sided pits that dotted the fields around Colkirk. He said that men with basket-laden donkeys dug out the marl to spread over the light soil. That was probably true, but the pits were much older than Lake. The marl may also have been used to build wattle and daub cottages for farm labourers, as many more would have been employed on the land in time past.

In fact, most houses had a pit or depression in the

soil next to them. Many of the pits were divided into two parts by a bank of earth, which suggests they were also used as stew ponds for fish. In my boyhood, many of them still harboured fish for we used to catch them. The early morning was the best time for fishing and I have vivid memories of jumping out of bed at sun-up, flinging on some clothes and dashing out into the dewy morning equipped with hope, rods, lines and a good supply of worms.

The roads we walked on, in fact all the roads in the district, were just grit and stones. The only difference between the high roads and the farm roads was that on the latter the stones were round and unbroken, while the main roads were repaired with broken flint and stones.

Old Howe, the roadman, spent all summer sitting at the roadside cracking flints with a long-handled hammer. He left neat piles of wedge-shaped flints at all strategic points. Then in winter he would go out with his barrow and tip loads of sharp flints into the waterlogged holes and ruts in the road, where they would lie until they were ground in by the iron-tyred wheels of carts and wagons. Everyone did their best to avoid them if possible, so it took most of the winter for the stones to work in. By that time another hole would be ground out by drivers avoiding the loose stones, so old Howe was never out of a job.

He had a son nicknamed 'Hoppy' Howe, as he was lame owing to a tubercular hip. This was long before the National Health Service, of course, but somehow or other Hoppy was sent to hospital. He was cured and restored to such health and strength that he was later accepted into the London police force. A number of village boys took the same employment.

The Black Death, the Reformation and sheep completely altered the face of Norfolk after the Middle Ages. The era of large estates and large farms came in. Another great change was wrought by Coke of Holkham and 'Turnip' Townsend with their intensive type of farming, which increased the population again. Hordes of labourers were required to till the soil and hoe the new-fangled swede turnips which would keep through the winter and feed the cattle. Another of our farm men, old Bartaby, told me that the men themselves almost lived on swede turnips in the 'bad times'.

I do not remember seeing any evidence of strip farming around Colkirk, but there were a number of small pieces called 'pightles' round the village which had evidently been cultivated for hundreds of years. I once ploughed out a bullock shoe, for in the old days bullocks were shod like horses for work on the roads or for driving to market in the shafts of a bullock cart. Another time, on the point of a share, I unearthed a gold wedding ring with an ancient inscription on the inside.

In the barn one day I found an iron rod with a cross piece at one end and a pear-shaped bulb at the other. Old Burton told me that it was a wheat dibble. The land was marked out in lines by a wooden marker (we used a similar tool for marking straight lines when baulking for mangels or swedes) and a man made holes all along the line with a dibble. A woman followed him with her apron full of seed wheat and dropped seeds into the holes – 'never less than two and never more than three'.

I can remember horse-drawn wheat hoes, about ten V-shaped hoes mounted on a bar. Hence the need to drill straight. That was in the days before selective weed killer – men were forced to rely on the skill of hand and eye.

There were two ruined churches within a mile of Colkirk and the meadows round them were scarred with mounds and ridges, evidently the remains of houses. I have never heard their history but I have read that the Black Death hit Norfolk very hard. Even Lake's race memory did not reach back as far as that – at least if it did he never mentioned it. I wish I had asked him; but it's too late now.

A wheat dibble

CHAPTER FOUR

THE FERRET AND THE FOXHOUND

We boys made friends with some professional rabbit catchers of the district and spent many a long day with them in the winter months, when they were at work within walking distance. We would find out in advance what farm they were on and when they saw us in the distance they would wave their spades to guide us to the spot. They always seemed glad to have us with them. We tried not to make a nuisance of ourselves and did as we were told – 'Quiet boy!' – and when the moment came, we helped out with the digging. In return they showed us the mysteries of their art.

We would approach the selected warren very quietly, taking particular care not to tramp hard with our feet, for noise travels a long way underground. Rabbits warn others of danger by thumping with their feet, and when the ground is hard, and honeycombed with 'workings' the sound can carry many yards, both underground and in the open, warning all outliers to take cover.

Having reached the warren, the first thing to do is to

put purse nets over the holes. Rabbits are very cunning in the way they place their bolt holes. The exits are often hidden in a clump of nettles or low bushes where, even if you can spot them, it is impossible to arrange a net. A bolt hole is a clean, round hole with no spoil scraped out to give the place away – unlike the main 'front doors' and normal exits. A man with a gun would be placed in a commanding position covering these bolt holes that were difficult to net, and a dog who knew his job would be sent to some tricky place where it was not possible to shoot. Our companions were not out for sport – cartridges cost money.

It is a strange thing, but the weather seems to affect rabbits' inclination to bolt. On some days they would rely on the network of runs below to escape the fierce, creeping ferret, rather than face the dangers above. Often one of the rabbiters would remark, 'They o'nt bolt much ter day. Wrong wind.' And he would be right. But whatever the weather, there are always some outlying rabbits – you even find them in dry, frosty weather. If you know how to set about it, it is easy enough to pick up a rabbit quickly with a gun or a good dog.

Once the bolt holes were looked after, it was time to put the ferrets in, several of them 'coped' (that is, muzzled) to prevent them making a kill underground. We were told to keep back and stand still. There would be a few thumps and rumbles underground and then suddenly, without warning, a rabbit would be bundled up in one of the nets. Someone went forward quietly, quickly dispatched the rabbit and reset the net. Bang would go a gun; or perhaps the dog, quivering with excitement in the place where he was told to wait, would dart forward and snap up a fleeing rabbit.

The excitement did not last long. Presently a disgruntled ferret would appear at the mouth of a hole, saying with every line of his body, 'What's the use? I can't kill with my mouth tied up!' Occasionally a coped ferret would stay in the warren, to be picked up during the digging operations or enticed out with a piece of smoked liver from a newly killed carcass. I have seen a ferret picked up after being lost several days. Sometimes, if he was still missing at the end of a day's rabbiting, a trap would have to be set for him – a ferret box with a heavy lid propped up on a piece of stick, to which was tied a rabbit's liver. During the night the ferret would come out, attracted by the smell of the liver and perhaps by the smell of his old box, and in the morning there he would be, curled up in the straw.

After getting all the rabbits that would bolt came the hard part of the job. A line ferret was put into the warren – that is, a ferret fitted with a collar to which a long line was tied, usually marked at intervals of a yard with a knot or a bit of coloured rag. The job of the line ferret is to show where the rest of the rabbits are, and a clever one will herd them all into one part of the burrow and stay with them.

A line ferret is usually not muzzled and his reward is the blood of the victim he fastens on to. He will seize the neck of the hindmost rabbit, biting into the jugular vein and sucking its blood. Cruel? Well nature is cruel.

It is one of the hard facts of life that we live by the death of others. Even a strict vegetarian eats his cabbage by the death of slugs, snails and myriad insects. This is a tale of country life, of people who live close to the real things of the earth.

The digging, which is the hard work of rabbiting, is made easier by skill and experience. One can learn a lot by listening to the scuffles and bumps going on in the warren below and thus get an idea of the best place to dig. The run follows no particular pattern – a burrowing rabbit will turn aside at a stone, a root, or just a bit of extra hard ground. This is where you need knowledge of the ways of rabbits. The knots show, perhaps, that eight yards of line have been taken. So you cut a trench a tad or two out, hoping to break into the hole. If you are lucky, there is the line. You then lie on your stomach to find which way the hole is going, and you dig again, and so on until the end. I have seen as many as a dozen rabbits taken out of the 'room' at the end – or it may be just one mangled corpse that is unsaleable.

Finally, it is an unbreakable rule that diggings *must* be filled in.

After a morning of this, how splendid to sit round a clear fire of dry wood in a sheltered spot and tuck into a meal of bread and cheese! No one worries about rabbity hands and earthy fingers. A boy's paradise! Then it is 'Come on boys!' and we are off again, gathering up nets and ferret boxes to take them to our next stand, where the whole process is repeated.

Presently, as the red ball of the sun dipped towards

home, there would be a 'Hulloo!' in the distance, which we would answer, and a pony cart would appear on the rutted farm road to take away the catch. I can see now the green patch of grass on the frosty ground, where the warm furry bodies lay. We helped to load the cart, the dog was whistled up from some private business of his own (probably connected with rabbit innards) and then we said 'Good-night' to the rabbit catchers and set off across country for home and tea.

Tired and cold? Not a bit of it, and we should have had little sympathy if we were. Today I am old and civilised and might perhaps be a little squeamish about the blood and stink and the rough ways of our companions. But at the time it seemed all right, and perhaps gave us a certain basic toughness which stood us in good stead in the trenches, not so many years later.

Our experience that winter fired us with ambition to own a ferret ourselves. By next autumn we had saved enough to buy one, together with the necessary equipment. We made a house for him, and a ferret box to keep him warm and dry when he was not working – this is essential, for a ferret that is cold and wet won't work. We had to get a male, or 'Jack' ferret – our rabbiting friends had no inhibitions about explaining the reasons why we must not buy a female. It was typical of the sort of people they were that they would not sell us a ferret themselves. No, we must bid for one in the market on our own. They recognized that this was an important part of the adventure.

So we went off to market one day, with all our money in our pockets and a stout bag to carry our purchase home. On the way down we saw a great trail of smoke across the horizon. Some children we knew came by on the road, driving home, and they called out 'Hurry up! Baker's on fire!' We had a quick look, but were not to be diverted from our purpose. Hands on our pockets, we hurried on to the market.

The ferrets were in a line of wire-fronted cages, alongside the chickens and rabbits. We looked at every one. There was one big fellow with black tips to his yellow fur. He looked clean and happy, and had a nice face. That was the one! We did not budge from that lot, not even for a second look at the fire, until the auctioneer got down to it. Jack was knocked down to us for five shillings – a lot of money in 1903, but he made much more than that for us in the months that followed.

Our business done, we went to look at the fire. Billy Baker's shop was well and truly alight. The fire had started in the paint shop, it was said, and there didn't look to be much hope of saving the place. It was a gun shop, besides selling paint and tools, and there were wild tales of Billy rolling out a barrel of gunpowder amidst the smoke and flame. Whether this was true or not I can't say, but people did load their own cartridges in those days. I had bought loose gunpowder there myself. The volunteer Fire Brigade worked gallantly, but the nearest water supply was the river some distance away.

The town pump was used to supplement the supply – an old manual fire engine was dragged to up to it and volunteers pumped the handle in turn. When the trough was full the fire engine crew took over, working up and down two long rails on either side of the engine. The result was a thin stream of water descending spasmodically into the fire. Presently the entire front of the house fell into the street, blocking it with rubble. Nevertheless, the firemen did manage to stop the fire spreading to the other shops crowded along the narrow street. Only the shop immediately opposite was scorched.

Our ferret was a great success. He soon became very tame and we used to play with him as one might play with a cat or a dog, but only when he was not being worked. On rabbiting days it was strictly business. He only once tried to bite me, when I happened to be wearing a woollen glove he had not seen before and did not recognize by smell. We quickly cleared the rabbits from our small place – though they soon came back into the desirable residences, empty and to let. Then we entered into a contract with a neighbouring farmer, the father of some boys we knew.

On that farm there were no big warrens, which in any case we could not have tackled with only one ferret. The rabbits lay in small burrows, some no deeper than an arm's length, but few and far between. It would not have paid professional rabbiters, as they need big sandy warrens to earn a good day's pay. There were three hundred acres to the farm, which gave us plenty of room to move about and more rabbits than you might expect – indeed, we were to get all the rabbits we could on this farm, and the farmer gave us sixpence each for them.

They were generous terms. We were soon bringing in a dozen or so at a time, and so Jack was quickly paid for.

*A meet at Colkirk in the 1920s.
(The author's aunt, Kate Chambers, is in the middle)*

Another of our winter sports was following the local hunt, if it met anywhere within sporting distance. The Meet itself was a stirring sight on a cold, sunny winter morning, with the Master and Huntsmen in pink and many of the riders also. The ladies wore trim black habits and white cravats. The younger women sported smart bowler hats, the older ones toppers. All of them rode side-saddle. I don't remember women riding astride until after the First World War – certainly not with this hunt, which hunted Royal country in the vicinity of

Sandringham. But we did not go only to see the Meet – we followed the hunt on foot.

There is more to fox hunting than a lot of people imagine. It requires deep knowledge of the countryside and the ways of foxes, and considerable skill and daring. The Meet itself was more of a spectacle and a social event, and it did not interest us boys very much. Often we guessed, or we might be told, where the first draw would be, and we would go straight there. It might be quite a way from the Meet, and to get there the hounds and the field travelled at a 'hound jog', about seven miles an hour, which put quite a strain on boys following on foot. Sometimes members of the field who knew us would invite us to hang on to a stirrup and give us a tow.

Hounds were put into a wood or other cover (foxes, like rabbits, often lie out) and one of the huntsmen would tell us a good place to stand to see the fox break cover. We must be quiet until the fox got away and then yell like mad 'Gone away! Gone away!' The huntsmen would also tell us which way he was likely to run – according to the wind, the lie of the land and the earths, and other places of refuge.

We did not attempt to keep up with the field but cut across the country, for the fox nearly always runs in a circle. The rare occasion when a fox runs straight across country is an event, and is talked about in hunting circles for months afterwards.

From time to time the hounds will check, having lost the scent, and then they will have to nose around trying to find it, encouraged by strange cries from the

Whips (the huntsmen), who at such moments will call encouragement to individual hounds by name. We would take our cue from friendly Whips or from the more wily followers of the hunt, who would now edge off quietly in the direction they thought the fox would run. Thus we reckoned to be in at the kill.

The kill is not, in fact, such a bloody affair as many townsfolk seem to imagine. One snap across the back of the neck by the leading hound and it is the end of the run for the fox. I am sure it is untrue to think that the fox is in an agony of apprehension all through the run. Up to the last they expect to get away, and often do. I have seen a fox slipping quietly along a fence with hounds baying only a couple of fields away. He was intent on mixing his scent with that of a team of horses working in the field and turning at the headland.

It is these tricks a fox gets up to that make fox hunting so fascinating – you have to use your head and your eyes, and you need a deep knowledge of the countryside to keep on after your quarry. Hunting, as all true countrymen know, is not a question of a bunch of rich, bloodthirsty idlers running a poor little animal to death. Some of the field may be sufficiently well off to afford a second horse, brought along by a groom to a pre-arranged spot, but in many hunts this is the exception rather than the rule.

And a hard-working farmer, on his one day out in the week, is not going to ruin his only horse for the rest of the season by reckless riding. Indeed, sparing one's horse by thoughtful riding and an understanding of the terrain is one of the arts of hunting. It is the business of the

Master of Hounds and his huntsmen to keep up with hounds and be there at the kill, yet their horses seem fresh enough at the end of a run.

There are a few black spots. Occasionally one hears of a 'bagged' fox – that is, one that has been taken to a covert in a bag and let out at the appropriate time. This is not sporting, for the fox is confused and out of his country. It may happen when the Meet is held at the house of a landowner who also preserves pheasants. Foxes are apt to be scarce in such a situation, since a man who values his pheasants may not be above quietly taking a shotgun to his foxes! But hospitality requires that hounds find on the estate. Somebody does the necessary. The owner himself may know nothing about it.

We once saw another dark deed done. The fox had gone to earth. The field and hounds were kept back and a terrier was put in after him. Two small boys were allowed to stay and watch. The huntsman made a noose with his whip and held it over the mouth of the earth. Presently a fox popped out into the noose and the huntsman carried it away fifty yards or so, while hounds were kept back to give the proper 'law'. We went with the huntsman. Now I saw him do this – he took out his knife and nicked the fox's ear, causing drops of blood to fall on the ground so as to make the hounds' task easier. Then as he threw the fox down, he twisted its hind leg. The fox was 'chopped' very easily after that, of course, in a few hundred yards. Why the huntsman did it I don't know. We heard shortly afterwards that he had been sacked.

Foxhounds are friendly, almost clownish animals.

They would stray as far as they dared from the whippers-in to make friends with children who were drawn to the event. Sometimes hounds so far forget themselves as to go hunting rabbits on their own while a hunt is on. Sometimes when we were making for home in the late afternoon we would see a couple of hounds, looking thoroughly ashamed and hangdog, sneaking back to the kennels on their own. Often a mud-spattered horseman would pass us, too, with a friendly greeting. A sportsman will recognize another sportsman, however humble.

CHAPTER FIVE

GAMES AND DIVERSIONS

The home meadow, close behind the farm, was the scene of many of our activities. Only a few acres in extent, it was bounded by tall hedges. There were two huge walnut trees in the left-hand hedge and, at the bottom of the field, a towering oak and an elm. Looking out from under the walnut trees one could see miles and miles of country, a wide valley that ran up to the last ridge before the North Sea. How bitter the wind that howled across there in the winter! In snowy weather the snow always drifted against the west side of that valley and lay there long after the rest had melted.

The home meadow was our private domain. No village boys were allowed there. We shared it with the old pony and the hens. Occasionally the cows were left there for the night or the farm horses if they were wanted early in the morning.

During the winter it was the place for building snow huts and fortifications. Up in the corner of the stack yard there was a wood pile, a great heap of oak logs, and when we were old enough – far too young by modern standards – we would spend hours there with a cross-cut saw and an axe, sawing and splitting up wood.

Here also we made the bows and arrows, spears and boomerangs, which were our playthings. As the years turn around one lives these things again through one's grandchildren. A couple of summers ago I made a boomerang for a small grandchild in a rambling, partly wild Vicarage garden in Kent. We hunted through the shrubberies for a correctly shaped branch, cut it down, and whittled away at it laboriously with an axe, just as I had done in Norfolk sixty years before.

The meadow was also a grand place for flying kites. There always seemed to be a wind. We made our kites on the floor of the 'gig-house', which was our workshop and a favourite retreat on wet days. Like all the other farm buildings it had its distinctive smell –a mixture of harness, dusty pony cart seats, apples and slaughtered pigs. It was the place where the pigs' carcasses were hung up to cool after dressing.

We became quite good at making kites. By trial and error we evolved our own pattern which flew well and could be left unattended in a steady breeze. It was a great thrill to drive into town in the pony cart and look back to see our kite flying bravely over the treetops, getting smaller and smaller as we went down the lane.

In a hedge at the bottom of the meadow there was a hollow tree which was full of bees. Year after year we saw them flying in and out, and we often asked permission to try to get the honey; but this was denied. Then during the First World War I came home on leave from overseas. A returned hero was not to be denied, and that year the bees were robbed. Panels were cut in the hollow trunk at varying heights, and what a sight met our eyes when the first square piece was pulled out! The hollow was filled with comb, lying vertically, like a bookcase packed full of books. All we had to do was cut the comb, top and bottom, with a long knife, and pull out a great slab of it. We took well over a hundredweight of comb. It was surprising how many old ladies in the village immediately developed coughs and sore throats as the news got around!

Bees? There were none to be seen. It was bitterly cold frosty weather, and I think they had all retreated into the smaller hollow limbs of the tree for mutual warmth.

We had a cricket pitch in the meadow for home practice and occasionally we would invite guests to play with us. Then, as we got older, we made a tennis court. It was old turf and hard work very soon made it playable. Our dear grandmother often stood at an upstairs window and watched us playing in the meadow. Long after she died we would still see her there, in her black dress and white cap, her silver spectacle case gleaming at her waist. Were we frightened? Of course not. We could never be frightened of Grannie. It seemed natural for her still to be there watching us.

In the hard winters (in memory all the winters seem to have been hard) the church pit was a great place for winter sport. The same water that the village lads toppled into off the greasy pole during summer festivities would freeze solid. Over it we would make long slides, twenty or thirty feet long, which were not despised by grown up farm boys on clear moonlit nights.

We found some skates at home – the old fashioned wooden ones that screwed into the heel of one's boot. The village lads found some too – I expect one or two pairs came from the Rectory – and soon a group of us were floundering about, arms waving and legs slithering in all directions. After a while some of us – I think it was we farm boys, since we generally took the lead in such things – started to play ice hockey with rough stakes cut from the hedge. No sooner had we begun when out came one of the Rectory girls with a bundle of ash sticks for us.

We organized teams and, with coats for goal posts, played the game frantically and without heed for rules or falls. It is a good way to learn to skate – after a fashion! But one day a couple came out of the Rectory who really could skate. How we admired the grace and ease with which they cut figures on the ice! The pond cleared for them in a flash and we stood in an admiring circle to watch.

The pit was a heaven for boys at any season of the year. In spring and summer it was a favourite haunt of water hens. They used to shelter in the heavy thorn bushes which covered the steep banks and build their nests on low branches hanging over the water, sometimes floating on its surface. On a still summer evening you could hear the carp smacking with their rubbery lips under the lily pads. Some must have been of great size, but we could only catch the small ones. We could have tried night lines, but somehow it didn't seem to be quite playing the game.

The dogs loved hunting in the bushes along the water's edge, though what they were after I never discovered. Maybe it was only the water hens, who were perfectly safe, being able to swim, fly and dive, or it maybe the dogs picked up the scent of a stray otter in search of a meal of carp. The little river Wensum was not far away as the otter runs.

In the springtime the old pit would be alive with frogs, making a piping chorus as they went about their love affairs. Soon there would be a mass of spawn among the weeds, presently to hatch out into swarms of tadpoles. I remember passing the pit in early summer after a shower of rain and the whole shore and stony lane was covered with tiny frogs, all determinedly hopping along in the same direction, to the haven of wet grass and ditches that lay to one end of the pit. I suppose this could be the origin of the tales about showers of frogs.

The parish was full of pits. There was another on the far side of the village, which was also a favourite place with children. Several sorts of rushes grew there, tall and strong. One type was good for plaiting into baskets, whips and rosettes. The very tall, very straight ones made spears and arrows for mock battles. In summer, glittering dragon flies flashed across the muddy water and there were, of course, all sorts of creepy creatures to be dredged out.

One day, when we were playing there we heard a shriek from some girls nearby. Young Gerald, about two years old, had fallen in the water. He was buoyed up by shawls and petticoats, which little boys wore then, but face downwards. When I pulled him out he was black in the face, but he soon got his breath back and bawled heartily, so all was well. His elder sister, very much scared, said 'Don't say anything. I'll tell my Mum that he fell in a dyke.' And I never did.

We were always fond of exploring the countryside and we used to do this on the days when there was nothing going on, and no work for us to do, on the farm. On one such day, we set out to explore a segment of country then unknown to us. There were certain areas that we knew well enough to pinpoint every interesting pit, knowing whether it was filled with water or dry, and whether there were fish in it or not. Pits, which were dotted all over the Norfolk landscape, were the most rewarding places to explore.

This time we set off in our usual manner across country, scorning roads and taking our bearings from sun and wind. I can't say why we had never gone that way before – possibly because no one had ever spoken about it and we didn't know the farmers over that way. Being unknown country, it was well worth exploring. To this day I don't know exactly what direction we took, or how we came to the interesting place I shall describe, although we knew the way home – that is one of the first things a country boy learns.

We crossed empty fields, making for some clumps of thorny bushes that looked interesting. To get there we followed a faint track, such as one often finds in that part of the country – an ancient road, abandoned for centuries and now rarely used. We were not disappointed. At the end of the track we found something that was evidently an old moat, deep dug, enclosing an area of about an acre of perfectly smooth turf. The moat was overgrown with bushes and, in places, filled with very still, black water.

This time we did not go looking for moorhens' eggs or the possibilities of fish. Why not? Being keen fishermen, that was the thing we were always on the lookout for. And why didn't we rush into the green circle to stand on the place where people had lived hundreds of years ago? Normally that would have been the natural thing to do. But not here. The place frightened us, we didn't know why. The air was heavy with evil. The smiling green ditch that encircled the meadow held an invisible menace. Something seemed to be warning us – 'Do not set foot in the circle!' So we turned back towards home, leaving the ancient site unexplored.

What we could not understand was that we had never heard the village boys talking about the place. When we got home we asked several of the boys about it. Some said they knew nothing of it. Others just went dumb. One boy said he knew of it, but '…it's not good. Nobody don't ever go there. There's things.' That was all we could get out of him and we soon forgot about it.

There is a sequel to the story, however. Mrs Latterly, who used to help on butter making days, had a son called Ted who joined the Royal Marines when we were quite young. We used to look at him with awe when he came home on leave, resplendent in his red jacket. As the years passed by, he finished his time in the Royal Marines and came home, a cheerful bachelor, to live with his mother down the lane. Once when our house was full of visitors we were 'bedded out' in the cottage and he was very kind to us. I had toothache in the night and he brought me a peg of neat whisky – a soldier's remedy!

We grew up and the Great War came; and Ted Latterly was recalled to the colours at the same time as we joined the fray, together with the rest of the young men in the village. After an interval of five years those of us who were left came back to the village, Ted Latterly among them. This time he was definitely retired and lived a life of ease with his mother. Then one day Ted disappeared. He was no longer to be seen walking up to the pub of an evening in his blue demob suit.

After a few days, the young men of the village thought it was time to go and look for him. They seemed to know where to go – straight to the old moat, which we boys had found on that day when exploring the disused track. They found him there, dead. Nobody knows why he went there. It was miles from his usual haunts. Or was it that they knew and would not say? The Coroner brought in a verdict of 'found drowned'. Twenty years at sea and drowned in a puddle of black water.

I knew those boys well. Some of them came to work for me after the war. We often talked of old times and had a drink together after the day's work. But I never heard any of them mention Ted Latterly again.

On another day, we had won reluctant permission from Mr B— of Norton Hall for a day's fishing in his jealously preserved trout stream. He was really a kind-hearted man but having no children of his own he looked on boys with a suspicious eye – and with good reason.

We spent the evening before overhauling our tackle and making plans for the day ahead, so that come the morning we were off straight after breakfast. It was a glorious day of early summer. We had already reconnoitred the ground and it was about three miles across country to the point where we planned to start.

On closer inspection we found that the stream here was broad and shallow, not really suitable for our type of fishing, so we decided to walk on through the water meadows. The snipe were drumming overhead, far up in the sky, throwing themselves down in that reckless dive of theirs and making a noise rather like bleating goats. I believe the sound is caused by the wind vibrating through their stiff tail feathers. It certainly sounds like that.

Now and again as we walked our way downstream a couple of wild duck would go up with a startled quacking. They are shy creatures, but walking along very quietly on the soft turf we occasionally managed to get quite close before they took alarm. We got near enough to hear them chatting softly among themselves, a gentle conversational sound, probably about the caddis worms and other delicacies they were scooping up from the shallow parts near the banks. Or they may have been planning a new home in the crown of some old pollard willow overhanging the water.

How do the little ones get down to the water? I don't know, but they must go, and at an early age, for the old ones do not feed them in the nest. Rarely does one actually see young wild duck on the water. They are very clever at hiding themselves, and not without reason – there are so many enemies, both above and below. They can even be pulled down by large pike, although the big chaps would not venture into the shallow water in this particular spot.

The nests of wild duck are difficult to find. For one thing, the old mother duck sits very close and wears a perfect camouflage, speckled light and dark brown, quite different from her more gaudy mate. And the nests themselves are very carefully placed, sometimes in the rushes and trailing brambles at the water's edge, sometimes, as I have said, halfway up a tree.

Not so the water hen's nest, which shows up from the other side of the stream as an untidy lump of rushes and sticks, rather like an island of debris caught up in the last flood. But there is always water between the nest and the shore, and overhead cover from bushes protruding over the bank or a tangle of rushes. Most nesting birds are very careful about overhead cover, as their chief danger is from the air – rogue rooks, jackdaws, jays, hawks and magpies are always on the look-out for eggs or nestlings to steal.

So we walked along the bank as quietly as we could, looking for the underwater holes in which the fish lie. We were not interested in the odd trout lying in the shallow water, or out in mid-stream, but persevered in search of the quiet bends of deep water. At one point we were startled by a loud squawk, and a heron rose out of the shallow water and flapped sullenly away towards Square Wood, back home to the heronry. By the way he flew he must have had a full crop for his nestlings at the top of one of the tall trees. Herons do not seem to worry about cover. Probably they feel safe enough, with their size and their sharp beaks for defence.

We took a few roach from the holes during the morning, but the weather was too bright and the water too clear for much sport. We had some bread and cheese in our pockets – the bag was too fishy, even for a boy's stomach! – so we sat down in the sun to eat our lunch and wait. But the sun kept slanting down. After lunch we started again, working downstream, and in due course we came into the shadow of Square Wood.

At last! This was the place! The water was dark and deep, with little whirlpools at the bends and a rush of water out of them. Very quietly we stripped off the floats from the gut traces, bent on a small hook and impaled a small red worm, with the ends hanging naturally down. This was lowered into the water and worked gently with the current, slipping away downstream. Presently my line tightened. I was into a very nice trout, which *was* duly landed.

Then my brother, who always caught and lost more fish than I did, hooked a real big 'un. It rushed him across the stream, bang into a great bank of weeds, and there the old devil stayed. Nothing would shift him. we tried throwing clods of earth, sawing on the line, tugging from upstream and downstream. Still he sat there.

I told my brother to keep his line taut while I set off downstream to find a way across to the other bank. I found a single plank bridge after about half a mile, crawled across, and hurried back along the other bank to the scene of the action. There was the taut line, leading into a mass of weeds right under the bank. I found a branch and tried to slip it under the fish. There was a wriggle and a flash, and that was that. Nothing but the remains of my brother's trace in the water. Then I fell in.

The sun was going down, so we thought we had better make for home. It was a long way across country. By the time we came to the two walnut trees that marked the home meadow it was dusk. A rather annoyed Knights was catching the horse to go and look for us. He was the first person we had seen all day. Then our Aunt called out, 'Where have you boys been?'

I could see there was a storm brewing, so I quietly opened the bag and showed her the beautiful little trout. 'Did you catch that? Come in to your supper!' All was forgiven. Our Aunt was a good sport – and Grannie had the trout, buttered and baked in the oven. My clothes had dried off a bit by then and if anyone noticed them, nothing was said. I put them on again next morning, to finish off the drying process. It had been a good day.

*Maria Chambers (Grannie) and Aunt Kate behind
The Cottage about 1910*

CHAPTER SIX

COCKLES, CHOCOLATE AND THE KITCHEN GARDEN

Nelson

Very few hawkers or itinerant salesmen came into the village, since it was isolated and far from any main road. One, however, came regularly, and that was Old Earle, who brought cockles, herrings and samphire in their season. Ringing a hand bell, he shouted 'Yarmouth herrings!', 'Cockles! Real Stukey blues!' and 'Fine samfire!'

Samphire was a sort of edible seaweed that grew in the marshes around the coast. We used to eat it boiled, sucking the green flesh off the hard twiggy stems, and it had a pleasant sea taste.

'Stukey blues' were cockles from the coastline near Stiffkey. They were much prized and had a blue tinge to their shells, probably caused by the peaty water that drained off the marshes, and this must also have had something to do with their fat tastiness. I have seen women attired in short skirts and thick black stockings raking them by the bushel out of the muddy flats at low water. Donkeys and ponies waited patiently on the nearest 'hard' until it was time to carry home the results of the day's work. The cockles were then boiled in huge coppers which were standard equipment in many of the cottages. The same coppers were used to cook the cargoes of shrimps, mussels and whelks the men brought home in their boats.

Old Earle ('old' in Norfolk denotes affection, notoriety and general usefulness to the community as well as mere age) was a drunken old rascal. As Colkirk was about the end of his beat, there he often stayed until he was too drunk even to walk. Some kindly person would then heave him into his cart and the pony would take him home on its own, a distance of about ten miles. That way of getting home was by no means uncommon in those days. There were no motor cars to worry about.

There is one immortal tale about Old Earle that is told among Norfolk people to this day. It concerns a certain evening when the pony brought him home, drunk as usual, and his poor wife took one look at him and decided she had had enough of it.

'Look here, bor,' she said, 'I ha' had enough. We gotta part.'

'All right,' said Earle, 'we'll part everything.'

He picked up a plate, broke it in two halves, and said 'There's your half. Here's moin.' Then he proceeded to do the same with every piece of crockery he could lay his hands on. Finally he reached for the much prized clock from the high mantelpiece and turned to her saying, 'Which half will yer have, mornen or arternoon?'

There were two shops in the village. One of them was dignified by the name of 'Co-operative Stores', although it was generally known by the name of the person who was running the shop at the time. The 'Co-op' was a local effort, not affiliated to any national organization. Like so many things in Colkirk, it was started by the Rector, a man with very advanced ideas for the age.

The 'Stores' stocked all sorts of things, from clothing and boats to bacon – a sort of forerunner of the modern supermarket. It had quite a distinctive smell of its own – a compound of bacon, butter, cheese and corduroy trousers. We didn't often patronise the place, except for a time when we discovered that the new girl in charge was selling first class chocolates at an absurdly cheap price; but like most good things, that young lady did not last.

The other shop was owned by old Mrs Bailey, of whom we were all very fond. She had a fascinating little piece of gold let into her cheek, just below one eye. We never found out why, and we were too polite to ask. She sold all sorts of things and was famous for her pork sausages. When a pig was killed she generally had half of it. A man would carry the limp carcass round to her shop.

This was hardly the modern idea of a shop, being one of the cottages belonging to the farm, with a single room downstairs partitioned off. One side of the shop was mostly counter, the other was living quarters. All sorts of things were displayed in the tiny window, which was her showcase. We spent our Saturday pennies here, on wonderful sweets with a motto printed on them, pear drops, ha'penny bars of chocolate and chocolate drops. The chocolate had rather a queer taste, but it was sweet and cheap.

All the sweets were apt to taste of pork, as they were kept in flat wooden trays cheek by jowl with the meat and sausages. Also, the old lady didn't do much by way of washing. Still, we could make a feast of two ha'penny bars of chocolate and a pen'orth of broken biscuits. These came as a kind of surprise packet, a collection of little pieces of all sorts of biscuit, on the soft side and a bit musty, but good enough for a boys' feast. We would carry them off and eat them by ourselves in some secret hiding place.

For a time there was a third shop, but we never went in it for fear of the old lady who ran it. She looked like a witch, with gingery hair and penetrating eyes. I don't think any of the village children brought their sweets from her – like us, they were frightened. Her son and his wife lived with her for a while, but pretty soon they moved to another house in the village. It was never really a shop, just a little house with a few jars in the window.

Our childish fears of the old lady had a strange sequel. After this witch-like lady had died or gone away, the house was taken for a few years by a Captain Hoare, who used to live there, stabling his horses with us, when he came to Colkirk in the winter for the hunting. Then two dear old ladies took it and they lived there for years. While they were there, some relatives of theirs from South Africa came to visit them, bringing a small boy, about four years old, who had never been in England before.

While his elders were in the sitting room he went running about the house exploring, as children will. They heard his feet pounding up the stairs and into a room overhead. Silence; then the footsteps pattered out of that room at full speed and down the stairs. The child burst into the room and ran up to his mother saying, 'Mummy! Mummy! I don't like that old lady upstairs. She made faces at me!' There was, of course, nobody there.

Another house in the village was reputed to be haunted and that was the farmhouse (The Grange, opposite the school) which my parents took just after the First World War. I returned there when the army had finished with me, and it was the first home to which I brought my newly wedded wife.

The house itself was not very old, about eighteenth century, but there was an interesting curved pond in the garden and a continuation of it in the adjoining meadow, suggesting a moat and the site of a much older house. We never saw anything, but the doors had a habit of opening quietly in the middle of the night. Perhaps there were 'things' I could not see, not being born in the chime hours.

My brother might have done better, for it was he who saw the early nineteenth century man in the bedroom we shared as boys in my grandmother's house. He described in the most positive detail the blue tailcoat and brass buttons at the back, which we identified many years later as the costume of that period. I slept in the same room but was conscious only of an occasional, nameless fear. I saw nothing. That was as it should be, according to the laws of the supernatural and the prophecies of Mrs Burton, our 'white witch', at the time of my brother's birth.

The kitchen garden was looked after by a succession of old men, some of them industrious, others rather more bumbling and lazy. I suppose it was really a sort of old age pension for them. Anyway, the kitchen garden was always well cultivated. It was entered through an arched door in the garden wall, the top half of the door being of vertical iron bars.

Of all the old men who did the garden I remember Nelson best. He had a kindly face, a saintly face, with a straggly bush of whiskers all round it. He always wore a black felt hat, perfectly round in shape, with a black cord around the crown – not unlike a cardinal's hat. Only old men wore these hats and it was a puzzle where they bought them. One day I had the cheek to ask Nelson where he bought it. He paused, looking at me for a moment with his old eyes, and answered, 'At market.' He rarely spoke to us, just went on steadily with his digging, yet he is the one I remember.

The garden door had to be kept shut, for cows passed it on their way between the road and the meadows. Indeed, it was impressed upon us from the earliest age that gates and doors were to be shut. A countryman's habits die hard – even now, living in a suburban road, I find myself instinctively going to shut other people's garden gates when there is no cause to go through them.

This particular door held a fascination for us. It led to all sorts of delights, mostly of a gastronomical order – gooseberries, currants, strawberries, green peas, apples, and filbert nuts; for there were nut bushes at the end of the garden. We were free of all these treats, provided we did not overdo our raids. If we got stomach ache that was our own business – we need expect neither scolding nor sympathy.

At home in the evenings our time was very often spent in reading boys' adventure books, for we learnt to read with enjoyment at an early age, much earlier than children seem to now. From these books we gained inspiration for all sorts of larks. For example, we read in one of our books about the making of a pitfall to catch elephants and other wild beasts. Our imagination was fired. What might we not catch with a pit in the kitchen garden? So we started digging. The soil was light and easily moved and we soon had a nice, deep hole.

Carefully following instructions, we then covered it with thin sticks, then broad leaves, then finally earth. It looked suitably innocent and we awaited our quarry. Next day Aunt Kate fell into it. After that, pitfalls were out.

In one corner of the kitchen garden, close by the greenhouse and cucumber frame, was a well and a pump. Here we used to arrange swimming races with frogs and toads in the tub under the spout. The pump was very old and would not work at all unless one poured water down the lead barrel of the pump. Later it was dismantled and a well head with windlass and bucket was substituted for the old pump, which put an end to our frog races as we had to promise not to open the well. I think it must be true that handling toads can cause warts because our hands were covered with them, but they all disappeared after the pump went.

The pipe that went down the well from the old pump was made with great oak logs with a hole bored down the middle of them. Each log was carefully fitted into the next to make a continuous pipe. When the pump was dismantled the logs were still sound and they were put to use as gateposts.

Of all the good things grown in the kitchen garden, the cucumbers claimed the most attention. There was a tremendous to-do about them. As summer approached, two loads of muck were brought round from the yards and dumped in the garden. This had to be left to heat and then it was turned over and made into a square heap, a bit larger than the cucumber frames. A few barrow loads of earth and the cucumbers were planted, with the frames placed on top. But that was not the end of the story. Every evening the frames had to be watered, every morning aired. As new manure was required for this operation every year, the rest of the garden benefited.

The greenhouse was an old-fashioned lean-to affair with a small fireplace outside and small, narrow panes of glass, but it was remarkably efficient. There were grape vines inside and a miscellany of potted flowers. We did not often go inside, although there were no rules against it.

CHAPTER SEVEN

SUNDRY AND MANIFOLD CHANGES

The village playground, other than the lanes and hedgerows, was known as the 'Campyng Land'. This was not, as you might imagine, a place where people pitched tents or gypsies parked their caravans, but the place where the old game of 'campyng' was once played. It was stopped by law during the eighteenth century because so many people got hurt playing it – at least, so it was claimed; but no doubt many saw it at the time as yet another expression of the enslavement of country populations through the 'modernization' of agriculture by the powers of the day, who were greedily fencing and gathering the land into large farms and estates.

So many old customs, of which I saw just a few lingering remains, were choked to death at that time by the new system of large scale farming – created at the expense of common land and small holdings – which then employed, at a miserable wage, the men whose free living it had destroyed. They may have been a bit better off than they had been when scratching their living from the strips and pightles around the village, but they lost their freedom.

Campyng was a sort of primitive hockey, played with hedgerow clubs and a wooden puck. Two villages were matched against one another and the battle raged anywhere between them, with no rules and no boundaries, along lanes and tracks and on any bit of open land that came to hand. It raged fiercest on the piece of ground known as the Campyng Land. Bearing in mind what happens in spite of rules and referees in certain of our present day games, one can easily imagine heads and limbs getting smashed when a village band was defending the honour of its village on home ground.

When I was a boy we played cricket on that same land. Even this sometimes ended in bloody noses – inflicted on one occasion by myself. I was supposed to be cricket captain that year and a hulking great youngster disobeyed my orders in the course of a match. I immediately set about him. He was heavier than I was but it was not a fair contest – I had been taught to box at school by an old army sergeant who believed that the best form of defence was attack. However, we remained friends afterwards.

An old man in the village told me a story about a former Rector who wanted to enlarge the churchyard and decided to annexe a sizeable slice of the Campyng Land. I don't know if he had a legal right to do so, but in the past people did not always bother too much about the strict legality of their actions. The men of the village rose up in revolt at this desecration of their Campyng Land and there was something approaching a riot. This must have been quite a surprise to the Rector, for even in my day the old man who told the story spoke with awe at the idea of such defiance of ecclesiastical authority. The Rector, however, was no coward. He faced the crowd as they shouted and shook their fists on the other side of his fence, and the authority of the Church proved too much for them. They just melted away. Now people lie buried in what was once part of the Campyng Land.

The last piece of common land to be taken under the Enclosure Act was at Common End, where nevertheless the name lingered on. The people who lived there when I was a boy were considered a wild and rebellious lot. The chapel was there, too – another indication of free thinking and lawlessness! The annexing of this bit of common land happened long before I was born.

All the landowners, including my family, had a share of it. Our share was too far away to be farmed comfortably, so it was let. I remember how the whisky decanter used to be brought out when the tenant came once a year to pay his rent. After I learnt how the land came to us I felt uncomfortable about it, even as a boy. But it is all the same now. The land is no longer ours and much of Common End is built over with little bungalows.

Our farm consisted of a number of small pieces of land scattered all over the village, all of it in fact built up of pightles or peasants' strips enclosed under the

Enclosure Act. My grandfather came to live in Colkirk as a sick man having given up control of West Wood farm in the neighbouring parish of Great Ryburgh. The Chambers – my mother's side of the family – were an acquisitive lot in the past, but the art petered out before my generation. I gathered from my mother that they kept up considerable state in the old days, when servants were cheap and the price of corn was high.

Local feeling for the old landowning families ran pretty deep when I was young, even among those who had served their interests for the pitifully low wages that were paid in those days. I once had occasion to take a pony and trap to be done up – to match the little mare that I bought shortly after my marriage. I had hardly got into the yard when the proprietor spotted me and walked over, saying decisively, 'You are a Chambers!'

'Well, my mother was,' I said.

'Same thing,' he replied. 'I was dog boy to your grandfather'.

He made a very good job of the trap, being a craftsman of the old order – no cheap or shoddy short cuts and three good coats of varnish. Last time I went that way the yard and the sheds were gone and a couple of new houses stood in their place. The twentieth century deals hard with the traditional crafts.

The old chap told me, without a trace of resentment, how my grandfather used to keep a bundle of hazel rods in the stables, and any farm boy who did wrong met with summary justice on the spot – or rather, on his behind! How I wish I had encouraged him to tell me more, but at that time I was living very much in the present and its day to day problems. I had no idea (and who else did?) what a gap the coming slump and a Second World War were about to create in our lives.

My mother grew up on the farm at Ryburgh, but for some reason she rarely spoke about it. Once she told me about the harvest supper they used to have with all the men in the big barn, to celebrate when all the crops were in. She recalled how they stood her on the table to receive the toast of 'The Family'. The evening's sing-song always started with two old men, standing one each side of the trestle table, singing verse and response:

> *'I sing the green-oh!'*
> *'Who is the green-oh?'*

The song is written down and listed in books of folk music today, but as those old farm men almost certainly could not read they must have learnt it by word of mouth from their fathers and grandfathers. There is a theory that this kind of doggerel is really corrupted Latin, originally taught by the priests in Roman Catholic times. The songs certainly have a liturgical flavour about them. If this is so, they did well to survive in our Cromwell-ridden county!

Colkirk parish church survived the religious upheavals almost intact. Only the niches of the saints are empty, possibly smashed by the Puritans, or diplomatically removed by a village priest. It must have been a very embarrassing time for any priest who started in the time of Henry VIII and lived through Mary's reign to Elizabeth's; or again to live from the reign of Charles I through Cromwell to the Restoration. Much depended, of course, on the politics of the local Lord of the Manor and the kind of protection he could give. But I expect the work of haysel and harvest went on just the same.

There is an interesting little window in the church, now blocked up, which gave on to the sanctuary. It was believed to be a lepers' window, but I wonder if it could have been a window to a holy woman's cell. There is a similar window in a church near Norwich which is fully authenticated. In medieval times it appears to have been customary for a 'holy woman' to build herself a wooden shack against a church wall and never leave it. She would mend the church linen and vestments, do odd jobs for women in the parish, and give her advice to those who sought it. She might also have dabbled in a bit of magic, such as curing warts and brewing love potions. In return she was kept in food. The window enabled her to hear Mass in her own hut.

Our church had a solid looking square tower with a peal of bells which was considered unsafe to ring. Old Goodman, the sexton and clerk, used to toll three at one time, with a rope in each hand and one tied to a foot, producing quite a pleasant 'Ding dong ding!'

Just to write this conjures up memories of warm, sunny Sunday mornings, sitting in the family pew with a clean collar on and looking up at the outline of the Rector's chin showing through his rather scraggy beard as he stood in the pulpit where the sun streamed in. The rest of the day was our own, provided we kept reasonably clean and didn't make too much noise.

The last time I went to a service at Colkirk Church, maybe thirty years later, it was all just the same. The bells still went 'Ding dong ding!' – rung by Bert Goodman's successor, with his two hands and one foot. He remembered me with a grin as I came in through the South door, which passes under the belfry.

Bert was a great cricketer in his time, a demon bowler and a bold batsman. It was a joy to see him open his shoulders and deal with a loose ball. He had shoulders about a yard across and arras like young oak trees. As a boy I played for the men's team once or twice when they were short handed and very proud I was. In my first away match they put me in first, no doubt on the principle of using the expendables first to test the bowling. To my great surprise – and everyone else's – I made five.

After the match we retired to the pub on the edge of the green, where the men had their beer and I had my lemonade. Then came the sing-song. Man after man got up and sang unaccompanied a long sad song about love and death and the field of glory – the theme never seemed to vary. It was the same men who went off to war not so long after; and me with them. The Norfolk's took it hard and few of them came back. The names of the dead are listed in the church. More unusual, so are those of us who came back – a sort of roll of honour.

My brother and I enrolled in the Norfolk Yeomanry on 4th August 1914, the first day of the war. In 1919, when the survivors were demobbed, I did not think to be back in khaki in twenty years' time. But so it was. It was not for anything as abstract as the national cause – more, it seemed, one ought to do something to save old Colkirk from being overrun.

Nevertheless, it was – by hordes of American airmen and by people from the nearby market town who had done quite well while the war was on and thought it would be nice to live in a quiet village. As a result, Colkirk ceased to be a true village and became a mere sleeping place for town workers. I suppose it was inevitable – part of the general trend of things. But it is horrible to see the place you love drained of its virile blood and bereft of its soul.

Enrolling in the Norfolk Yeomanry on 4 August 1914 in Fakenham. Alfred and his brother Leonard are among the mounted horsemen. Photograph by May Bone

CHAPTER EIGHT

BARLEY

Harvest

Harvest time is, of course, the natural climax of the countryman's year. It is the time he gathers the fruits of his work and of his land – a busy time, a festive time, and often an anxious time. On the work and the weather during these few weeks depends the feeding of his livestock, the basic economy of his farm, and ultimately his family, throughout the year. So important and deep rooted is this sense of dependence on the harvest that even in the cities, where the impact of the seasons is scarcely felt, it retains its hold on our language and folklore, and the rituals of the church.

In a rural village a century ago, this was not only a matter of folklore and tradition – it was of immediate and vital importance, involving the whole village.

In our part of Norfolk, and indeed throughout the county, barley was the staple crop. Wheat and oats were also grown, employing the traditional four course rotation, but the rather light soil favoured barley. Everyone aimed at a 'malting sample', which fetched a higher price but at the same time needed much greater care in harvesting and was therefore more expensive to produce.

A bad harvest could halve the price of malting barley – a 'bad harvest' was either a wet one, or else one with a dry period falling just as the grain was beginning to plump up. What the maltster wanted was grain of a uniform bright yellow on the outside which, bitten in half, was plump and white within. If a sample of barley offered for malting contained one 'grown out' grain (already starting to sprout), that was enough – the sample would be handed back without a word. 'Steely' barley with a hard, clear kernel – the result of dry weather – was also rejected.

At the maltings, grain was spread out on a warmed floor and thoroughly wetted; and it was important that it should all sprout at the same time, for it is the sprouted barley that makes the malt. A generous nature turns the starch of the kernel into sugar as it sprouts. Hence the importance of an even sample. Grains that have sprouted too soon will lie and rot on the malting floor, while the steely ones take too long to sprout.

All this was known to masters and men alike and it was almost an act of religious faith to treat the barley tenderly – a faith which owed at least part of its strength to their liking for good beer!

In my earliest recollections, before the First World War, barley was still harvested with scythes. I have seen as many as twelve men move in echelon across a twenty acre field, led by the 'Lord of the Harvest', each cutting his four feet and laying the 'swathe' in a neat row behind him.

The swing of the scythe took an upward direction at the end of the stroke, thus leaving slightly higher stubble for the barley to lay on. Each man had his 'rub' or whetstone, carried at the small of his back in a specially made pouch, which was threaded on his belt. Every now and then he would straighten his back, pull out his rub and give the blade of his scythe a few strokes to keep a keen edge on it. But not too often – everyone had to keep pace with the Lord of the Harvest and none must fall behind. It was he who set the pace, bargained about wages and settled any disputes that arose among the men.

Once cut, the barley was watched and tended as carefully as a newborn babe. A little rain would do no harm, indeed it was said to 'mellow' the grain, but once rain had fallen it must not be left wet. Everyone turned out after rain to 'lift' and, if necessary, turn the barley where the force of the water had beaten it down. Frequent turning was also necessary if there was much green stuff among the corn as this would heat in the stack and ruin the malting qualities of the corn. Once dry, the corn was gathered into sheaves and these were balanced against each other in 'shocks', or groups, of three, spaced so that the wagon could pass between them.

The harvest involved everyone living on the farm. The women took food up to their men in the fields to save precious time. There were two main breaks in the day, 'elevenses' and 'fourses'. Everyone would break off and gather in the shade of the trees to rest and consume their 'whittles' – mostly chunks of fat bacon, cheese and bread, washed down with cold tea. There was also a traditional 'harvest cake', a sweetened bread dough made with yeast and with a few currants in it. The children came along with their mothers. It made a merry party.

The older boys who were still at school used to 'take' a harvest in the summer holidays, committing themselves to help out on a particular farm and making themselves generally useful. They had their wages at the end of harvest, like the men. The Lord of the Harvest would arrange, on request, to draw an agreed sum each week which would then be deducted from the lump sum at the end. This applied to men and boys alike.

No overtime was paid, but of course it was to the men's advantage as well as the master's to work until dark and get the harvest over as quickly as possible. If the harvest turned out long and wet, the Lord would eventually go to the master and say, 'Look, Maister, we can't go on no longer with this hare wet.'

A wise man would reply, 'All right. Ye go on weekly wages as from next Saturday.'

The last harvest wages I paid on my own farm amounted to a weekly £12 and that was in a wet harvest.

The scything of barley in Norfolk ended with the First World War. After that it became impossible, due to the shortage and expense of labour, and farmers took to cutting and binding their barley with a 'self binder'. This machine had already been in use for a long time cutting wheat and oats. Also, when I was a boy, there were still a few old 'sail reapers' about, which cut the corn and scraped it off the reaping platform in bunches, to be tied by hand.

We had a binder and being, in those days of smaller farms, more thrifty of the corn we harvested, we would prepare a field for work by mowing round the headland to make a path for the binder. A large field would, in the same way, be divided into sections for easier work. Today the approach is quite different – with their huge combine harvesters they just pile into the bowed and bent corn, because it is cheaper to waste grain than to pay for picking it up. Sound farming economics, no doubt, but we would have considered it sacrilege. The corn that grew round the hedges, and even that which had been 'laid' in places by wind and rain, was mowed and carefully gathered into sheaves.

That was a useful job for us boys to do. I learnt the method very young. You shuffle along the ground until there is enough corn lying across your feet to make a sheaf. Then you grab a handful of straw in each hand, just below the corn ears, and with a quick twist the ears are tangled together. This forms a knot which will hold when it is passed under the sheaf. You then deal with the two loose ends of your bond, twisting them together and tucking one end under the bond. The other end is left standing up. I can still see the man who taught me to do this, holding the completed sheaf up by the bond to show me how well the bond stood the strain. It sounds laborious, but an expert could do this almost at walking pace.

To me in those days the self binder was a wonderful piece of work – in spite of its tendency to jam and break down for a host of different reasons, all of which had to be understood and avoided if one was to make any headway with it. If the tension of the knotter was too tight the sheaves would be too tightly bound and, sooner or later, the twine would break – a fact that will be readily understood by anyone who has threaded a non-automatic sewing machine.

If the spring of the sheaf release mechanism was not tight enough, a great oversize sheaf would be rammed tightly together and its girth would prove too bulky for the needle (a great curved piece of iron) to get round it – result, another jam-up. Or a badly spun length of twine may seize up a knotter – another hazard. The cut corn fell on to a moving platform of canvas; but if there was a heavy dew on the corn, the wetness on the canvas would tighten it up, so it would stop going round and the corn would not be carried up to the binding and releasing mechanism.

I graduated to driving a 'binder' once I had learnt the ins and outs of the machinery, but I must confess there were many occasions when it got the better of me and I never felt I was a hundred per cent master of the machine. All its foibles had to be kept under surveillance while perched high up on an iron seat driving three horses – and they, too, had their tricky ways.

They soon learnt that if they pulled out a bit from the line of corn their work was easier, and if they pulled in too much towards it some would be missed. To keep up the work, and cut the corn correctly was tough labour and a continuous strain on the poor beasts. There was no let up either for them or for us. But one could not afford to be sorry, even for oneself. Least of all for oneself. There is a sense of urgency about harvest which even horses feel.

That is the delight and the pain of farm work, going back to the beginning of time. One must work to eat – get in the fruits of labour in time, or starve through the coming winter. It is the natural law. Sometimes I think things are too easy in these modern times. We may imagine we have conquered Nature, but history shows that Nature has a habit of coming again into her own. Civilisation as we know it is balanced on a knife edge and not nearly as stable as it appears. If and when the crash comes, it will be the primitive man, the man of the soil, who survives. It has always been so.

The sheaves of cut corn were not carted in right away, leaving a clear field such as you get now within a day or two of cutting. They were 'shocked', that is, stood up on their butt ends with their ears inclining inward, making a reverse 'V'. Six or eight sheaves went into one shock. A well made shock would stand a lot of wind and rain, but it was best not to let barley stand too long if it was intended for the maltsters, as the outside corn weathered more than the inner and that tended to make for an uneven sample.

The price of barley fluctuated violently in some years and the less reputable dealers gambled on the price. One year a man came to the harvest field where we were carting barley. There was a stack half up and it was beautiful barley, bright and well ripened. He took a look at it and then offered me a good price for my whole crop, half of which was still in the field.

Something struck a warning note. I replied that he could have it, but I wanted a written guarantee of the price and acceptance. Rather reluctantly, he gave it. Almost that very day the weather broke. We got the rest of the crop in a fair condition, but not as good as the sample he had seen. Now the maltsters were worried about the quality of the barley on British farms and had begun to place orders abroad. So down came the price of barley. I threshed out my corn in due course and set off with a sample to the Corn Hall, where I ran my man to earth and said, 'Here is your barley. Where do you want me to deliver it?'

He replied, 'Sorry, old chap. The price has gone

down. I can't take it.'

'Oh yes you will,' I said. 'Here is your written guarantee.' And he had to take it. He had gambled and lost. If the price had gone up he would have held me to the bargain. But if one wanted to play safe, the answer was to stick to the well known dealers one saw in the market each week. Not that they were angels, but they knew that a shady deal would not pay because it would spoil their market.

However, I have known a well known horse dealer employ an underling to do his dirty work. Indeed, I was once taken for a ride myself in this fashion. It was just after I had taken on some extra land and I needed another working horse. I made my needs known at market and soon a young fellow approached me saying that he worked for Mr W——, a dealer with whom I had dealt before, and he had a horse that would just suit me.

He did not take me, as was usual, to one of the two hotels in the market square, where private rooms were provided, complete with pens and ink, for the benefit of people who wanted to conclude a deal. Instead he led the way to an obscure pub just outside the town. That raised my suspicions for a start. However, the horse looked all right and I said I would have it on trial.

'No,' he said. 'It's a good horse. Give me a cheque and it's yours.'

I said, 'All right, but I'll post date the cheque.'

Next day old Lake and I went to try him out in chains beside a steady old mare of mine. It went well enough for a round or two. Then the fun started. He reared. He kicked. He did all that a horse should not do. He was a rogue and no use to me. I sent my wife straight off to the bank to stop the cheque, while I set off at once to return the horse, a journey of some eight miles. Mr W— was not there, so I handed him over to the stable man saying, 'This is your boss's horse. I don't want him.' And walked home.

Next market day, when I ran into old W—, he adopted a very self-righteous air. I wouldn't have had it happen to you for anything!' he assured me. 'But you know these young 'uns. He's such a keen lad!'

I smiled to myself and thought 'you old hypocrite!'

CHAPTER NINE

BROTCHES, FRAILS AND GABBLES

Thatching

Thatching is a skilled job and men would go round the farms during and after harvest, taking on the thatching of the stacks, It was important for the job to be well done, because sometimes the stacks had to stand and be kept dry all winter.

I learnt to thatch and I suppose I could do it now,

given the right straw. The stuff the modern 'combine' leaves behind it, crumpled up in a bale, would be of no use. You need a load of good wheat straw, plenty of water and balls of binder twine, or tarred string if the stack has to stand a long time.

I have seen straw ropes made for thatching, twisted by a little crank with a hook at the end – I remember as a boy seeing one of these instruments hanging up in the barn and getting somebody to show me how to use it, but I never saw it seriously used for thatching.

The thatcher also needs a bundle or two of 'brotches' (Norfolk for 'brooches'), a term used for hazel rods split down the middle into halves or quarters. In Dorset, I discovered much later in my life, a much thinner hazel rod is used, soaked in water and twisted and bent into a hairpin shape to hold down the thatch. But I reckon that our north east winds, straight off the sea, needed something stronger.

The brotches held the thatch in place. They were rammed into the stack and walloped down with a heavy piece of wood attached to a handle. This instrument was also used for banging down the thatch, although a good piece of thatching should not need much banging down. A long ladder was essential, long enough to lie flat on the pitch of the roof.

At the start of thatching, the wheat straw is placed on the ground in a long heap and thoroughly wetted. Straw is then drawn out in handfuls and 'gabbled' against the feet and legs of the operator, the fingers drawn through the straw to straighten it out and to remove short pieces.

One would need five or six 'gabbles' to do one length of roof, about three feet wide. The 'frail' is then laid out on the ground – this is just two pieces of hazel rod, tied loosely together at one end, with a loop on the end of one rod and a notch at the end of the other.

Before starting, the thatcher has a look over the roof of the stack and fills in any hollows left by the stacker at harvest time. All is now ready for laying the first length of thatch.

The thatcher would seize a frail full of gabbles, hump it on his back and climb the ladder. Setting the frail half up the roof of the stack, secured by a brotch, he would take up the first gabble. This he would place with the thick end over the eaves, for two reasons. The thick end would make a bit of a 'hip' at the eaves, as one often sees on tiled roofs, and the thin end of the next gabble would blend into it. And so on up to the ridge, each one overlapping the one below.

One generally pinned down the exposed gable end of the stack with a criss-cross of twine and brotches. The rest was just tied top and bottom. Each gabble had to be re-tucked under along the side of the previous gabble. I once let an old man do the thatching – I was busy elsewhere – and when it came to threshing time there were long, vertical green lines all down the roof. He had not tucked in the edges of the gabbles. After that I did the thatching myself.

The top gabbles were laid projecting slightly, so when the other side was similarly thatched they would make a trim ridge. One sometimes stuck a brotch at the top of the gable end, wound round with straw. In the old days, I believe, elaborate 'dollies' were made to decorate the ends of the stacks. I imagine the custom went right back to pagan times, when our heathen forbears perhaps put up the effigy of a farm god to keep off evil spirits.

Later, they would be saints. It came to the same thing. Then the stern Norfolk non-conformists who supported Cromwell tried to erase the custom. But it lingered on to my young days. It had a utilitarian reason too, for the point of the gable would be vulnerable to the wind and a good strong brotch would hold it down. I remember once, after doing a real good job on a hay stack, feeling very tempted to finish the gable end with a 'dollie'.

As a working farmer I found it paid to do the thatching myself. I couldn't do it better than a professional thatcher, but I could certainly do it better than the odd labourer. I also thatched some of my sheds. Here the technique would be slightly different. The ends of the gabbles were tied down to the rafters and were made thicker.

I made a bullock shed and yard entirely of natural posts and rails, clad with furze faggots, tied strongly with wire. A neighbouring farmer was only too glad to get his furze on a rough meadow cut down. I made a warm, cosy yard and shed, which would not only keep cattle in but pigs too. I had a score of pigs in the yard at one time and no animal ever got out of it, not even my artful old cows.

CHAPTER TEN

THRESHING

The days during which the corn stacks were threshed formed one of the highlights of our winter. First came the excitement of seeing the threshing tackle arrive on the afternoon before work began. The threshing drum and the elevator were manoeuvred most cleverly by a great traction engine through the narrow lanes and gates. The huge single cylinder engine looked cumbersome, but responded marvellously to the slightest touch of the driver on the regulator which controlled it. That godlike man with blackened face stood like a Roman charioteer, balanced on the footplate between the furnace door and the coal tender.

It was the standing complaint of the farmers that the tackle always arrived with an empty tender and went away with a full one. But there was good reason for this; for it often had to travel a fair distance between one stand and the next and those iron monsters had a big appetite for fuel.

A day's threshing took the best part of a ton of steam coal. A lot of fuel would be used initially in 'setting' the drum and the elevator which carried the corn on a belt from the stack to drop it into the drum to be threshed. In soft, wet soil such as the winter too often provided underfoot, there was considerable danger of the heavy machinery becoming bogged down. The tackle had to be delicately eased into place, with the drum an exact distance from the engine, so that the driving belt which worked it from the engine would be neither too tight nor too slack. The drum also had to be the correct distance from the stack, neither too near nor too far away. Then the straw elevator had to be positioned exactly, as this in turn was driven by a belt from the drum.

The initial plan for all this would already have been worked out by the farmer and his men at harvest time and the stacks would be positioned accordingly. The farmer saw to it that his stacks were each the right size for a day's threshing – the usual size was about ten yards by five. Behind this lay a sound economic motive, for if nightfall left a couple of hours' work still to be done, a full day's work would be charged the next day.

The threshing tackle was manned by two workers, the driver and his mate. A gang of men usually followed it round and they were taken on by the farmer and paid by the day to help out with the work. Later on in life, with the responsibility of my own farm, I found it could be a hard decision to make early in the morning when the weather was not too promising. We had no weather forecasts in those days to help our judgement of the day ahead. Should one give the word to start and risk having to stop later on and cover everything up? It was an understood thing that once work had started it was a day's work, both for the men and for the threshing tackle, and the farmer paid accordingly.

On the other hand, if one said 'No threshing today!' they would scatter to do their gardens or some other odd job and there would be no threshing at all until the morrow. There were always a number of men within a radius of a few miles who were prepared to work that way. Some had smallholdings of their own which took only a part of their time. Some did hedging and ditching and other such part time work, opting for the freedom and hazards of casual work rather than full-time commitment to one employer.

In our boyhood, however, these considerations did not bother us. We would wake early, and seeing the frosty thatch being taken off the stack, we would know that it was a threshing day. Exciting things were happening. We would run out to see the engine driver opening the furnace dampers and chucking on lumps of coal to get a good head of steam. As we watched he would go over the engine with an oil can, carefully oiling each joint, and put his water hose in the iron water cart, which had been filled with water at the nearest pit and drawn alongside the engine.

He worked with the grave abstraction of a man who knows he is being watched. Moreover, he knew small boys, who were to be kept at a distance from moving machinery. He was one of the few men who would never let himself unbend to small boys – or anyone else for that matter. Was he not the god of the machine? The lord (and part owner) of fifty pounds per square inch of steam power? Was he not responsible that every belt and wheel was working properly; that nobody was caught in the moving machine; that the corn was coming through clean; that the boy whose job was to rake away the refuse from under the riddles, was keeping it clear and free to do its job; and that the straw was making progress up the elevator?

So there he stood on his rocking footplate, hand on the regulator, gravely watching for the first signs of trouble. He could tell if things were building up for a jam and quickly shut down the power before it got too bad. It might be that the 'feeder' (the man on the stack who forked the corn into the machine) was putting corn in too quickly, or the sack man had closed down his chute too much and grain was building up inside, or a belt was breaking, or broken, or coming off.

Any of these things could cause a jam-up and a half hour delay if they were not spotted quickly. But a quick adjustment, either human or mechanical, was usually enough to set the whole thing rattling and grinding into action again. The threshing drum itself was carefully sheeted up each night to keep out rain and dew, which could cause all sorts of trouble if it got into the works.

While the driver primed the engine for the day's work, his mate would be busy uncovering the drum and stacking the covers away while the odd men who helped them out by the day would be putting up the hinged working platforms, which were propped up with poles. As at harvest time, these day-to-day men usually appointed among themselves a 'Lord', who was then responsible for settling where each of them should work and who acted as a go-between in any discussion or argument with the farmer about wages and hours. The arrangement worked smoothly enough, indeed I never heard of any trouble arising from it.

Each man knew the job he was appointed to do and did it uncomplainingly. The 'feeder', who had the important task of feeding the corn into the drum, was skilled in that job and was more or less permanently assigned to it wherever the drum might be working around our district. If the corn was bound in sheaves the 'bond' had to be cut, then with a sideways motion the corn was spread evenly over the drum.

It was essential to have a reliable man on this of all jobs. If something went wrong and the sheaf was fed in whole, all kinds of disastrous things could happen. The feeder's hand, caught up in the bond, might get drawn into the machinery causing terrible damage to himself – this has happened. In any case, a whole sheaf could choke up the box, causing breakages of belts and gear. Even if it went through, the grain would not be properly knocked out of it and there would be complaints from the farmer.

Any farmer passing by could tell by the 'chug' of the threshing drum at work if the feeder was doing his job properly or not. The drum was a huge box-like implement, about the size and shape of a large furniture van, festooned with belts and pulleys on the outside, which could be seen going round as they operated the various riddles and fans within. The actual threshing implement, which could not be seen by a small curious onlooker on the ground, was a number of revolving beaters working at high speed.

The grain, straw and chaff passed through a series of riddles which gradually separated them out, so that eventually the three came out of different openings. Revolving fans, blowing a steady stream of air, fanned away the dust, chaff and the lighter weed seeds, which formed a cloud around the machine as it worked and eventually settled on the ground.

The heavier foreign seeds were riddled out and the clean grain poured out in a golden stream from a chute, where it was caught in a series of sacks. A man stood there in attendance, ready to shut down the chute when a sack was full, tie it up and quickly substitute another sack. Then the chute would be opened and once more the precious grain would come forth. It was beautiful, almost awe-inspiring to watch that final, perfect appearance of the product of all our labour through the farm's year.

At the other end of the machine the straw, now empty of its burden, was ejected on to the receiving end of the elevator, which carried it to the straw stack. Here again there would be a man, or men, catching and distributing the straw so that it packed together evenly into the stack, which would gradually get used up as bedding for the farm animals through the winter.

I am sometimes asked if we didn't use flails for threshing in my boyhood. The answer is no, I am not *that* old, although I have used a flail to knock out small quantities of corn – say for the fowls. A flail consists of two heavy sticks joined together by a leather hinge. One end is held in both hands and the other is banged on the corn as it lies on the floor in the barn. This is the reason for the huge barns still to be seen around the country, particularly on the wide, rich plains of East Anglia, which has always been corn growing country. The corn used to be carried into the barns at harvest time and stacked at each end, high up to the roof. Then it would be slowly threshed out with flails as it was needed through the winter.

I did know an old man whose memory went back that far. He told me how in his boyhood he would be awakened in the morning by the thud, thud, thud of the flails going in the barn before daylight. If it was not that, it would be his mother striking a light with steel, flint and tinder, to get the kitchen fire going.

With our threshing done in the open air, we wasted no time in carting the full sacks of corn off to the shelter of the barn. It must not be allowed to hang around and risk getting wet. If the corn was even slightly damp or sweaty ('clung' is the Norfolk word for it) it must not be left long in the sacks. A standard sack of barley weighed sixteen stone – of wheat, eighteen stone. As a young man I could carry a sack of corn on my back and place it where it was wanted, or if it was damp, 'shoot' it into a great yellow heap on the floor of the barn. Nor was this an unusual accomplishment – it was expected of youngsters grown up on a farm.

In my early youth the corn used to be 'dressed' before marketing. A hand-turned machine was used, rather like a miniature threshing machine, to sort through the grain a second time, blowing out any chaff that remained and any small and broken seeds, which would also detract from the quality of the sample. The machinery for 'dressing' was kept under cover and it was a job usually done on a wet day – one man would turn the handle to work it while another shovelled corn into the hopper at the top. The dressed corn came out of a chute to be sacked, finally, for market.

With the mechanical improvement of the threshing tackle the dressing machine was no longer necessary and stood idle in the corner of the barn. One year we hid a lot of apples in it. There was no need to hide them – they were not stolen. Boys like to have a secret hoard. But, like squirrels, we forgot about this one. Then one day in late spring Knights found the hoard and brought them to us. My, but they were good!

Finally, the day would arrive to take the barley into the great maltings, about three miles away. It was another highlight in the cycle of our year and we went along too, when we were allowed. We would watch as the wagon was loaded, then up we would go on top of the load. It was wonderful to lie up there on the sacks, the horses plodding along and the scenery going slowly by. From our high perch we could see over the hedges right down into the fields we passed, noting the kind and quality of the crops; or maybe a hare loping unhurriedly across a piece of pasture; or a covey of partridges squatting in the stubble. So much that was invisible to a person at ground level could be picked out from our exalted seat.

Then came the maltings – great brick buildings with huge, cowled ventilators poking out of the roofs. We might have to wait, if a wagon had got there ahead of us. Our load would be checked and samples taken by a clerk from the office, to be checked with the sample sold. Then it was the turn of the wagon to be unloaded. It was drawn under a sort of white penthouse, stuck on the side of the high wall. There was a trap door in the floor of the penthouse which projected directly over our heads. Presently there would be a shout from above and a bright chain with a ring on the end would come snaking down. Quickly the ring was turned into a loop and slipped over the tied mouth of the first sack. 'Stand clear!' And up the sack would go, swinging over our heads. We never saw the dusty, godlike creature up aloft who worked the controls.

Old Mr Wright, who kept The Star public house hard by and also owned the Smithy, once told me that in his boyhood the corn wagons started out for the maltings early in the morning, long before light. The team men would pull into The Star yard for a pint or two of ale with their breakfast and he would be hauled out of bed to serve them. He also told me how corn used to get lost on the way to the maltings, stolen on the road. The thieves provided themselves with a strong rope to which a sharp hook was attached. This would be slung from the branch of a tree which hung over the road. When a corn wagon came along the thief would nip up on the back of it, under cover of darkness, and stick his hook into one of the sacks. The wagon went rumbling on with the driver, no doubt half asleep in front, and the hooked sack would swing clear.

On the lonely road to the mills was an old, old oak, long since dead, and for some reason horses often shied at it. Could it be that one of the old corn stealers was caught and hanged there, in the harsh old days, on the scene of his crime?

CHAPTER ELEVEN

ANCIENT CUSTOMS, PLEASANT PASTIMES

The Barn

In the Colkirk I knew people, particularly the children, still observed old customs which had been passed from generation to generation, right back from pagan times. Some had been taken over, it is true, by the Christian faith, yet there is evidence that the roots of these country customs go much deeper and older than that.

Some of them involved the collection of pennies on

certain days of the year, which was itself an incentive for the children to remember the date! It is still so with Guy Fawkes Day, one of the few old traditions observed by children today. Nevertheless, I believe that the urge to observe old customs had, and still has, deeper motives than the collection of pennies. There was something rather pagan about the earnestness of the simple ceremonies as conducted by children – a sort of dedication of themselves. If one asked them why they were doing it, the reply would be 'we ha' allus done it!'

At harvest time, during the rest and refreshment of 'elevenses' and 'fourses', the boys would make posies for their buttonholes or caps from straws with the ears still on. I remember being taught this art with great earnestness and care. It seemed almost a religious rite, as indeed I think it was, as we sat quietly in the shade of the hedge with the marvellous harvest field smells all around us – the smell of the bright straw; the smell of the horses as they stamped and swished their tails close by; the smell of Stockholm tar mixed with wagon grease; the smell of dust and herbage and of men at work in the heat – not an unpleasant smell at all, in such surroundings! The straw posies we made had a special name, though for the life of me I can't remember it.

Christmas, oddly enough, was not such a good time for customary rituals in our part of the country – beyond the fact that just about everyone came to church and just about everyone ate too much! Carol singing was not so popular as in the towns and suburbs. The local church might have done something, had the Rector or the choirmaster been so inclined. But from what I remember of the elder members of the choir I can't imagine them going out carol singing, and the boys could not do much without their support. As for those small, tuneless groups of carol singers that appear on town doorsteps during the festive season, we just didn't get around to that sort of thing. Perhaps we were too proud, knew each other too well, to ask money for such slender value. Or perhaps we were just lazy – in the country it is a long way between houses!

The next festival was 14th February, St Valentine's Day. This was a great day for the boys of the village, who started out in groups to go round the houses singing – 'Good morrow, Valentine. You be the giver. We be the taker!' Thus they collected funds.

The real fun began after dark. Bogus parcels were made up, containing all sorts of rubbish. The custom was to place these parcels on people's doorsteps, rap the door, and then hide and watch the householder collect his fine looking parcel and take it inside. A variation was to tie a long piece of string to the parcel and snatch it away as he stooped to pick it up. Door knockers were a rarity in the village, but with the odd house that had one, the boys would tie a piece of string to the knocker and retire to a strategic position, pulling the string to produce a ghost knocking.

Turnip lanterns were another tradition of St Valentine's Day. A turnip, hollowed out and carved with mouth, nose and eyes, was made to look as much like a grinning skull or a hideous face as possible. With a stump of candle lit inside, this was supposed to frighten respectable citizens out of their wits. The point was that the grown-ups, of course, knew all about these things and played up to them. It was the boys' day and, within reason, the boys could have their fun.

All Fools' Day, April 1st, still lingers on, though I believe the rules are not so strictly observed as once they were. One had to be very careful about taking a message or answering a question on April 1st, at least until noon. After that you could go on being a fool in your own way, but it was no longer the thing for anyone to try to catch you out. If someone did try it and succeed, you could turn round and say – 'Ya! April fool yourself. It's after 12 o'clock!'

St Mark's Day, April 5th, was the time to cast spells. 'Things' walked on St Mark's night. If a girl wanted to know who her future lover would be, she must look into her mirror by candlelight at midnight and, after repeating some doggerel verse as an incantation, her true love would appear, looking over her shoulder.

Or, she must go into the garden at midnight and repeat a spell, causing a similar apparition to appear. He probably would, but he would be hot flesh and blood. Trial marriages are no new thing; in fact they were almost usual in the villages in those days. The Parson would marry them in good time and they would be happy and faithful to each other for the rest of their lives. Nobody bothered.

May Day, May 1st, was the day for little girls. A doll was obtained, or made from rags, dressed up in the last baby's Christening outfit and displayed in a linen basket, surrounded with wildflowers. This was taken round to everyone's door and shown as the Queen of the May. Donations were expected – and received. There was no nonsense of dancing round the Maypole, in fact I never heard of it in Norfolk. The Puritan parsons must have had their way there – or perhaps it belongs to another tradition which never took root in this predominantly Saxon part of England.

The characteristics of Norfolk people and their speech remain predominantly Saxon, in fact the actual sounds of words used by folk in rural Norfolk have their echoes way back in the history of our language. I once tried to read Chaucer in the original. At first I could make nothing of it. Then I hit on the idea of reading it aloud in broad Norfolk. Reading thus by the sounds of words, a lot of the sense became clear. It is a link that has held for 500 years, through the speech of the common people.

SOME NORFOLK DIALECT WORDS

Bin	*Manger*
Bish-a-barnabee	*Ladybird*
Blar	*To cry*
Bor	*Neighbour*
Buskins	*Gaiters*
Clung	*Damp*
Cop	*To catch*

Dickey	*Donkey*
Dodman	*Snail*
Dwill	*Floor cloth*
Hake	*hook*
Helve	*Handle*
Hol	*Ditch*
Hull	*Throw*
Loke	*Lane*
Mawther	*Young girl*
Muck crome	*Muck rake*
Pightle	*Small field*
Pishmire	*Ant*
Poke	*Sack bag*
Quicks	*Couch grass*
Redweed	*Poppy*
Roke	*Mist*
Shanny	*Silly*
Shuck	*To shell peas*
Wa'rmint	*Rogue*

The First World War put a strain on the link – the second broke it. Now we are all being reduced to one dialect – television talk. This is one of the reasons why I am using an unaccustomed medium and writing down my memories of the very last period in the history of rural England – trying to preserve at least a little of those times for my sons, for their children, and anyone else who may want to know something of the soil whence they sprung.

After May, there was really too much to do in farm and garden for the village children to indulge in any concerted games except cricket in the evenings. There was birds' nesting and stick whittling on Sunday afternoons. And during the hot summer days before harvest, the boys would go bathing in one of the pits, their sisters sitting sedately on the bank to watch. Bathing costumes were not worn and the sun was their towel. The favourite place near our village was Moss Pit, being shallow without any deep holes.

Cricket was the all important game, round which centred the tribal feeling and social life of the men and boys of the village. It was cricket that we played against other villages in the neighbourhood. I don't remember the boys playing many other games, though there was a year when football became quite popular, played on Saturday afternoons and moonlit nights on one of our meadows, known as the Star pasture. It was started by an enlightened school board, of which my aunt was a member. Somebody gave a set of goal posts. We had fun but never got round to playing regular matches.

After harvest, when the green acorns appeared, was the time for pop guns. One selected a piece of elder wood, straight and of several years' growth, and cut off a ten inch length. The pith must then be pushed out – a laborious job – and a piston had to be made out of deal, purloined from old George Nelson's yard. If it was the right sort of wood, straight-grained and tough, it would whittle down until it fitted neatly in the bore of the pop gun. Then, with plenty of spittle and much banging against a brick wall, a fuzzy end could be made at one end of the piston.

At the other end a neat handle was carved. All this would take many hours, indeed the activity would usually spread out over several days. When all was done, a green acorn was rammed into one end of the bore, using the carved handle of the piston as a hammer. Insert the piston, well lubricated with spittle, into the bore, and give it a sharp push – the result should be a satisfying bang. I feel sure this gave more pleasure than boys get from the elaborate toy guns of today, so easily bought from the shops. Probably the reason is that we had to work at our armaments, and had the satisfaction of knowing that our own skill produced them.

CHAPTER TWELVE
COLKIRK CRAFTSMEN

The Forge

The life of a village in those days depended on its craftsmen in a way which modern communications and technology have changed beyond measure. When something needed doing or when something went wrong, you were dependent on the skill of the village expert to put it right. So my tales of Colkirk end with an account of three men who fulfilled crucial roles in the village as I first knew it, one of them living on there to see his craft outdated and made redundant, a victim of progress.

George Nelson was the carpenter and village undertaker. Old George could make anything in wood, from a coffin (solid oak or elm slabs) to a five barred gate.

George and his son, young George, were to be seen sawing out boards in his sawpit – George perched on a plank on top and the boy in the pit below. A long cross-handled ripsaw was used and it was a skilful job, for the saw must not wander away from a straight line.

The sawpit was in a bit of meadow, away from the carpenter's shop. His fat old pony grazed there. It was a handy place, too, to leave his rough timber to season, being at the junction of two roads where a timber drug could be unloaded. His workshop was next to his house at the bottom of Hall Lane, a wonderful place which we regarded with awe and affection. It gave us the same sort of feeling as going to church, except that the smell was nicer.

There one could distinguish the sharp rich smell of pine, the heavy smell of sawn oak, and the distinctive smell of ash which when it is sawn smells like a hedgerow in spring. (In passing, it is one of the tragedies of a pipe-smoking old age that one loses the keen sense of smell which is part of the memory of childhood.)

The workshop was ankle-deep in shavings and devoid of any machinery except for a mortise cutting machine, worked by a crank which was turned by hand. Old George Nelson was very proud of it – I think it had been installed by his father in the early Victorian iron age. There was also a lathe, made entirely of wood – headstock, tailstock and bed, all of solid oak. On it, George turned up ash pitchfork handles (the grain must be straight), table legs, well-head drums and anything else round that was required. It was powered by a long treadle and strong legs.

When my father retired as an old man, he wanted a new barrow to push about the garden, so he went to George Nelson and asked him to make one. George wanted to make him a proper one – ash frame, oak bottom and elm sides.

'No,' said my father, 'it will be too heavy and, besides, will I last fifty years? Make me one of deal. That will last me out.'

'Deal!' said George, 'I can't do that!' However, under persuasion he did. It nearly broke the old chap's heart and for long afterwards he would grumble and mutter under his breath to me about 'yer father's barra'.

George had a long, thin nose and when he wanted a certain sort of wood for a job he would poke about among the stacks of planks looking, in his white apron, like an industrious bird hunting for insects in a hedgerow. He changed his outward appearance for funerals but could not disguise what he really was, a craftsman.

Arthur Goodman was a master bricklayer and also the Parish Clerk. When somebody died, he tolled the church bell, one stroke for every year of life. As most lived to a ripe old age, it was a lengthy business. People would count the tolls and say, 'That's old so-and-so. I knew he (or she) was failing.'

Being the clerk, Goodman would make the responses in church in a loud, hoarse voice – he looked the part too, a big heavy man with side whiskers. His pew was at the back of the church, graced with a large bible and prayer book, and besides making the responses he kept order among the young people who tended to congregate at the back of the church. He had very 'Low Church' ideas and I think it was largely because of him that the choir boys wore no cassocks, only little short white surplices. Corduroy knickers, black stockings and heavy boots looked a bit queer below the surplices, but cassocks smacked of Popery and therefore of the Devil.

Old Goodman was more than a bricklayer. He built a house for himself in his late middle age, and his daughter, a very old lady, was still living there when one of our sons visited the village some years ago. Two of his daughters, Maude and her sister Martha, came to us as nursemaids when I was a child, one after the other. We were very little at the time but I can remember them both. Maude we particularly loved. It was a large family and I expect the girls had plenty of practice bringing up their small brothers and sisters.

It was one of the brothers that later I pulled out of a pond. He had so many clothes on that he floated, but unfortunately face down. He was black in the face when I got him out, but soon he started roaring like a bull, which reassured me that there was nothing much wrong with him. I promised not to tell and never have, until now. I don't think it matters much after more than sixty years. Goodman was very good at building open fireplaces for burning wood, in a way that made them draw well. Faced with such a task, his favourite expression was 'That'll all ha' ter come out!' And out it came. He believed in building from the bottom.

Alfred c.1910

Frank Wright was the village blacksmith, shoesmith and wheelwright. I remember his forge as early as anything, for when still very small we were given rides to the forge when the horses needed shoeing. Later on we were entrusted with the horses by ourselves. We loved watching him make a shoe, hammering the glowing metal until, as if by magic, it was suddenly recognizable as a horseshoe. Then while it was still glowing it would be 'burnt out' against the horse's hoof, making a smell not so agreeable as some country smells.

It did not hurt the horse – the hoof, like our nails, is insensitive – but sometimes the horse would get nervous and object. The really tricky part was nailing the shoes in place. The nails had to go in at exactly the right angle or it would pinch or, worse still, prick the sensitive part of the foot. And there had to be enough nail to clinch properly – that is very important. To have someone returning to the forge with a nearly new shoe in his hand, and leading the horse that cast it, is not pleasant for either party. I can only remember it happening once. I led the horse in and handed Frank the shoe. Nothing was said.

Frank would never let us blow up the fire. It is an art in itself, not to be lightly undertaken. When we offered to work the bellows Frank would say 'Ye'll blow the fire all to bits!'

Looking back now, that is exactly what would have happened. A sudden blast, blowing hot cinders about, would have left a hollow in the fire, so that the iron would be cooling in one part and almost melting in another.

The bellows were huge, about five feet across, crouching in the shadows like a great toad. Its back was harnessed to an eight foot pole with a bullock's horn for a handle. Frank would heave on the handle with a steady motion, turning and twisting the iron held in the tongs and patting the coal round the job with a spade-ended poker. Heating a shoe was a comparatively simple job. Welding required much more skill – the two parts had to be heated to the right temperature and transferred to the anvil quickly, where they were then beaten into one by taps with a light hammer. There were no mechanical aids to determine the temperature or condition of the metal – only long experience – for Frank's father was a smith and his father before him. The very name 'Wright' shows that the family trade went back into history a long way.

To see Frank tempering a piece of steel was a joy. It must be heated to a bright red, quenched in a trough of water in front of the forge fire, quickly scrubbed bright with a piece of soft stone and then, when the band of colours running down the job had reached the right tint required for the temper needed, quenched finally.

Some heavy jobs needed a striker. The glowing red part was held on the anvil by a pair of tongs in Frank's left hand and a light hammer in his right. His striker would stand ready with a long-handled seven pound hammer poised to strike. A slight twist of the tongs, and down would come Frank's hammer on the spot where he wanted the metal to be hit by a smashing blow of the striker's heavy hammer. So a merry tune would ring out – twist, bip, bang – twist, bip, bang! It had to be done quickly before the metal cooled and it was wonderful to see the job taking shape under the quick hammer blows. (Later in life I was to see the heavy crank shaft of a big marine engine bashed out by a steam hammer which shook the ground at every stroke, but it was not so thrilling as seeing the village smith delicately shaping a farm implement.)

Talking of farm implements, a plough share is not forged. It is an easily replaced piece of cast iron. The plough coulter is forged, or it used to be, out of an iron bar. The coulter is the tool that slices through the soil in front of the plough breast, thus making a clean furrow.

The cleverest of all the things Frank could do was tyring a wheel. First, a strip of iron – about four inches wide if it was for a wagon – had to be cut the exact length of the circumference of the wheel, allowing a bit for the welding. One can imagine how difficult this would be, for the diameter of the tyre had to be just a little less than that of the wheel because it was to be shrunk on.

To that end a huge wood fire would be built in the yard, to heat up the tyre evenly and all at once. Meanwhile, the wooden wheel was placed on a circular metal plate with a hole in the middle to receive the hub. When the great moment arrived, the hot tyre was lifted out of the fire with long tongs and quickly placed over the wooden wheel, then quenched with buckets of water held in readiness. The tyre was now firmly shrunk on and would remain so until the wooden fellowes rotted. Carts with wood and iron wheels are still seen up and down the country, mostly cast aside, a few preserved in museums. Today they are rarely, if ever, used.

Frank Wright was not a really good shoesmith. Horses did not like him, which made them more restive than they should have been, and the fact that Frank suffered from a 'queer tummy' did not help matters. But he loved to make things, forging them on the anvil and drilling holes on his much-prized drilling machine, turned with a hand wheel. He preferred to punch holes in red hot metal if it was possible. He knew all about iron and had no use for the new-fangled rolled steel – it wouldn't 'work', he complained, and it rusted so easily. Also he didn't like what he called 'round nuts', meaning hexagonal ones, but preferred the old-fashioned square nuts, which fitted his home-made spanners.

Yet he would seize on any new idea and would make anything to a rough drawing or just a description of what one wanted. When I went into business as a farmer myself he made me a number of implements and tools, among them a water cart, and he put into effect an idea of mine for an implement to fix to a plough, which broke up the 'hardpan' at the bottom of the furrow. This needed an extra horse, but allowed the sugar beet to get through and make good roots. Another job he did which impressed me very much was to make plungers for water pumps. They were made of a piece of elm clad with leather with a leather clack valve, and they had to fit perfectly against the barrel of the pump. It was a craftsman's job indeed.

A year or two after the Second World War I went back to Colkirk to see the old place again and look up old friends, those who were still alive. I went to the forge to see Frank Wright. It was a tragic meeting. The shoeing stall was shut up with grass growing over the entrance. The forge fire was out and cold. The whole place was stacked with bits of old iron and inside was an old, bent man, standing forlornly by the once well kept drilling machine, now rusty and unused. Frank had lost his reason and just stood there, hands hanging down, muttering. I had to get out of the place quickly or else burst into tears.

I think this is the place to finish my tale of the passing of the old world. Whole generations of craftsmen, indeed a whole way of life, were thrust aside to make way for progress. I find myself wondering, was it necessary to kill off the old crafts, the old trades, in the process of making a bright, new, brittle world? What have we gained in losing a thousand years of patient craftsmanship? What has the countryside gained if it has lost its soul?

Alfred sketching, 1920s

Alfred in Wells, 1920s

ACKNOWLEDGEMENTS

This edition was originally transcribed and edited from Alfred's manuscripts by his niece Jenny Pearson, to whom we are greatly indebted.

Opposite: 'The cows coming in for milking at the back of the Cottage', painted by the author *c.*1910

This is roughly contemporary with the photograph of *Maria Chambers (Grannie) and Aunt Kate behind The Cottage,* the site of which is on the right of the painting, where the sheds are. The same bucket appears to be visible in both images